THE MATH TUTOR'S FRIEND

$\cos^2(\theta) + \sin^2(\theta)$

$2\pi = 360°$

trig indentities

reference angle

$\tan(\theta) = \frac{\sin(\theta)}{\cos(\theta)}$

DeMoivre's formula

BOOK 7

TRIG-ONOMETRY

by
Joe Cyr

The Math Tutor's Friend *Book 7: Trigonometry*

Cyr, Joseph

The Math Tutor's Friend / Book 7 – Trigonometry / by Joe Cyr

Third Edition ISBN: 978-1-967039-48-7 Joseph Cyr

Second Edition ISBN: 979-8-9879759-7-8 Joseph Cyr

First Edition ISBN: 9798476392590 Amazon Kindle, 2022

3 4 5 6 7 8 9

Subjects: Main category of book – *unit circle; trig ratios; trig functions; trig identities; polar coordinates; vectors – high-school math*

2.Other category - math tutor reference

Cover illustrations by Ramon E. Owen. Cover Design by Joe Cyr

Book design by Joe Cyr

The Math Tutor's Friend Series

Books

1 - Pre-algebra

2 - Algebra I

3 - Algebra II

4 - Algebra III

5 - Probability & Statistics

6 - Geometry

7 - Trigonometry

8 - Pre-calculus

9 - Series index & "FLASHMATH" handouts/reference cards

visit joe-cyr.com

Op. 0550

THE MATH TUTOR'S FRIEND

BOOK 7

TRIGONOMETRY

Sections

About the "Math Tutor's Friend" Series:

Teachers impart knowledge guided by lesson plans in a course syllabus. Math tutors, on the other hand, face varying topics at differing levels in successive sessions; for example, pre-algebra, advanced algebra, geometry, trigonometry, statistics, or even calculus! This series addresses the issue and is intended to provide the tutor with needed quick review of high-school math topics. The material is not presented rigorously nor does it abound with proofs and explanation that are available in textbooks - the tutor already knows this stuff - but a quick refresher is always a good move before diving into a tutoring session.

Too many in high-school have reached the 11th and 12th grades lacking basic math skills such as knowing the multiplication table (relying on a calculator to multiply 6 by 7, for example), handling signed numbers, and performing operations involving fractions - knowledge that should have been acquired by the fifth grade! I try to address this deficiency by devoting the first few tutoring sessions to these fundamentals and providing handouts of multiplication tables,etc.

Books 1 through 8 of this series include a number of quiz questions (answers explained) that the math tutor may find useful.

Book 9 is an overall index of topics. It contains a set of summary "FLASHMATH" pages for quick reference or for use as handouts.

This third edition corrects typos found in previous editions and expands discussion of some of the topics.

Happy Tutoring!

Joe Cyr

Sarasota, FL

●

THE MATH TUTOR'S FRIEND
$\cos^2(\theta) + \sin^2(\theta)$
$\tan(\theta) = \frac{\sin(\theta)}{\cos(\theta)}$
$2\pi = 360°$
reference angle
It's easy !!
quadrant II
degrees
radians
Pythagorean theorem
Mr. Math
THE UNIT CIRCLE

About Book 7: "Trigonometry"

Section 1, *The Unit Circle.*

Emphasis -:

- Standard position for an angle
- Trig ratios
- Reference angle
- Special angles & the Unit Circle
- Sign of cosines and sines in the Quadrants

Section 2, *Trig Functions.*

Emphasis -:

- Mnemonics for reciprocal trig ratios
- Law of Sines
- Law of Cosines
- Complex numbers
- Trig identities
- DeMoivre's Theorem

Section 3, *Polar Coordinates & Vectors.*

Emphasis -:

- Vector terminology
- Displacement vs distance
- Adding vectors
- Dot product
- Cross product

• •

1. THE UNIT CIRCLE
TABLE OF CONTENTS

●

1. THE UNIT CIRCLE

Angles

An angle is said to be in *standard position* if its vertex is located at the origin of the *x*-*y* Cartesian coordinate plane and one ray is on the positive *x*-axis. The ray on the *x*-axis is the *initial side* and the other ray is the *terminal side*.

Anatomy of an Angle in Standard Position

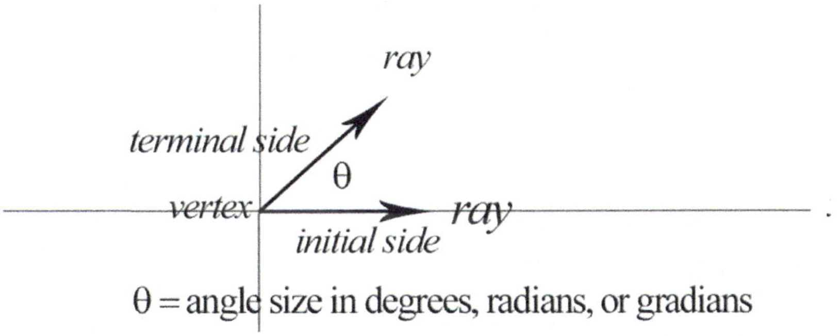

[This book addresses angle measurements in ***degrees*** or ***radians*** only.]

Trigonometric Ratios

Consider the *unit circle* shown to the right. It has a radius of one unit and an angle, θ (*theta*), in standard position, as shown. The right triangle formed within the circle is ΔAOF. Let's denote the sides by their end points. Using the angle θ as reference, it is clear that the unit radius of the circle, denoted by "OF," is the *hypotenuse* of the right triangle. Now we identify the two other sides of the triangle, namely OA and AF. Side OA is *adjacent* to angle θ and AF is the side *opposite* from angle θ. The ratios of pairs of these sides are the *trigonometric ratios* (*trig ratios*) that have special significance in trigonometry and named as follows. The trig ratio "***sine***" (abbreviated "*sin*" is defined* as

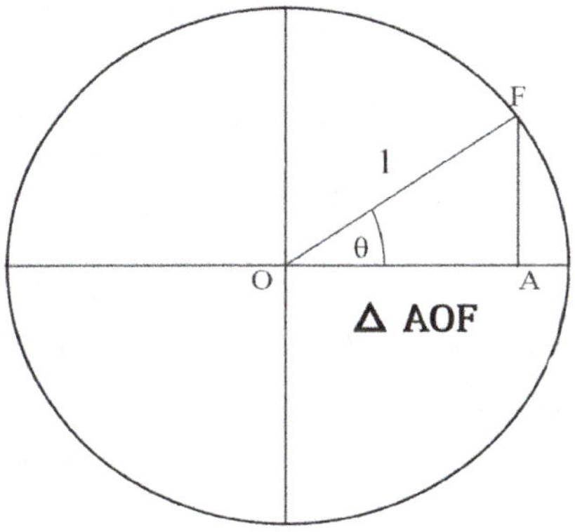

$$\sin(\theta) = \frac{opposite}{hypotenuse} = \frac{opposite}{1} = opposite = AF$$

and the trig ratio "***cosine***" (abbreviated "*cos*") is

$$\cos(\theta) = \frac{adjacent}{hypotenuse} = \frac{adjacent}{1} = adjacent = OA$$

* The definition with a "1" in the denominator is due to the unit circle; in general, for the sine, the definition is sine (θ) = ***opposite/hypotenuse***. This holds true for the oter ratios involving the hypotenuse.

Note that the (x, y) Cartesian coordinates of points A and F are $(\cos(\theta), 0)$ and $(\cos(\theta), \sin(\theta))$, respectively. Using the *Pythagorean Theorem*, we arrive at Equation (1-1), the *Pythagorean identity*: $\sin^2(\theta) + \cos^2(\theta) = 1^2$. Therefore,

$$\sin^2(\theta) + \cos^2(\theta) = 1 \qquad (1\text{-}1)$$

There are four additional trigonometric ratios to define:
tangent (abbreviated "tan"):

$$\tan(\theta) = \frac{opposite}{adjacent} = \frac{\sin(\theta)}{\cos(\theta)} = \frac{AF}{OA}$$

cotangent (abbreviated "cot")"

$$\cot(\theta) = \frac{adjacent}{opposite} = \frac{\cos(\theta)}{\sin(\theta)} = \frac{OA}{AF} = \frac{1}{\tan(\theta)}$$

Note that the tangent and cotangent are reciprocals of each other.
Now, the *secant* (abbreviated "sec"):

$$\sec(\theta) = \frac{hypothenuse}{adjacent} = \frac{1}{\cos(\theta)} = \frac{OF}{AF}$$

The secant is simply the reciprocal of the cosine.

Last, is the *cosecant* (abbreviated "csc"):

$$\csc(\theta) = \frac{hypothenuse}{opposite} = \frac{1}{\sin(\theta)} = \frac{AF}{OA}$$

The cosecant is the reciprocal of the sine.

Pythagorean identities:

$$\sin^2(\theta) + \cos^2(\theta) = 1 \qquad (1\text{-}1)$$

Let's derive two additional Pythagorean identities. First, divide all the terms of Equation (1) by $\cos^2(\theta)$:

$$\frac{sin^2(\theta)}{cos^2(\theta)} + \frac{cos^2(\theta)}{cos^2(\theta)} = \frac{1}{cos^2(\theta)}$$

Simplifying (use the definitions of tangent and secant),

$$\tan^2(\theta) + 1 = \sec^2(\theta) \qquad (1\text{-}2)$$

Now, divide Equation (1-1) by $\sin^2(\theta)$ to derive Equation (1-3):
From which

$$\frac{sin^2(\theta)}{sin^2(\theta)} + \frac{cos^2(\theta)}{sin^2(\theta)} = \frac{1}{sin^2(\theta)} = csc^2(\theta)$$

$$1 + \cot^2(\theta) = \csc^2(\theta) \qquad (1\text{-}3).$$

Just as it was used to illustrate the sine and cosine ratios. let's employ the unit circle to illustrate the tangent, cotangent, secant, and cosecant ratios. See the diagram below. Extend the line *OF* beyond the circle and construct a vertical line tangent to the circle at point *B* until it intersects the angled line at point *E*.

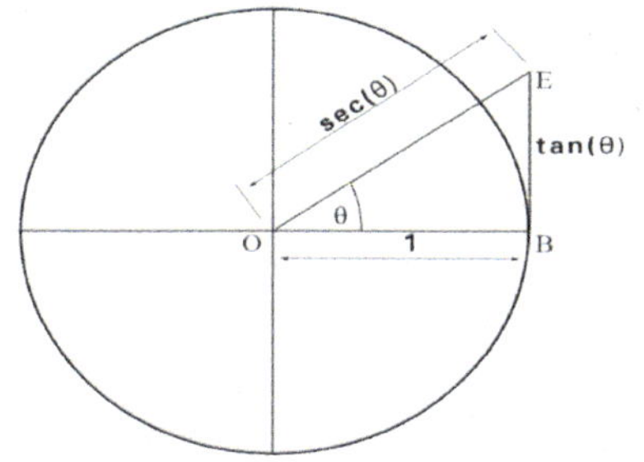

Δ OBE $1 + \tan^2\theta = \sec^2\theta$

This construction results in a new right triangle with base, OB of one unit (the radius of the unit circle), as shown. Applying the definition of tangent, we have

$$\tan(\theta) = \frac{opposite}{adjacent} = \frac{BE}{\mathrm{OB}} = \frac{BE}{1}$$

So the line segment *BE* must be tan(θ) as shown in the figure. Continuing with the definition of secant, we have

$$\sec(\theta) = \frac{hypothenuse}{adjacent} = \frac{OE}{\mathrm{OB}} = \frac{1}{\cos(\theta)}$$

Using the unit circle again, let's find the cotangent and cosecant. Extend the leg and hypotenuse of the previous triangle until their end points are exactly one unit apart, as shown in the figure below. We have a right triangle whose opposite leg has length of one unit. Determine the lengths of the base and hypotenuse. Applying the definition of cotangent, we have

$$\cot(\theta) = \frac{adjacent}{opposite} = \frac{OC}{\mathrm{CD}} = \frac{OC}{1}$$

So the line segment OC must be cot(θ) as shown in the figure.

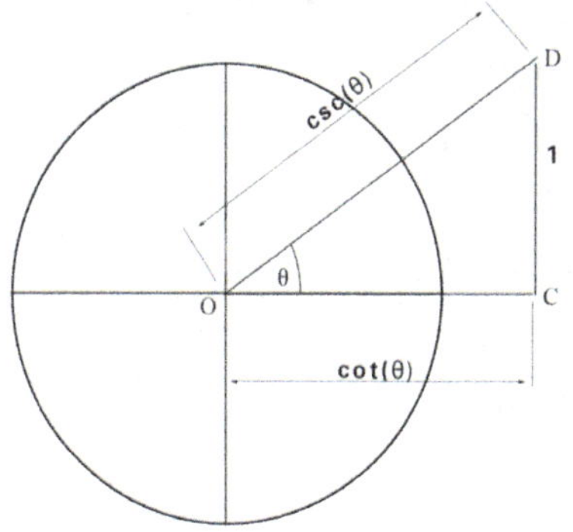

Δ OCD $\cot^2\theta + 1 = \csc^2\theta$

Continuing with the definition of cosecant, we have

$$\csc(\theta) = \frac{hypothenuse}{opposite} = \frac{OD}{DC} = \frac{OD}{1}$$

So the line segment OD must be csc(θ). The figure below summarizes everything discussed so far:

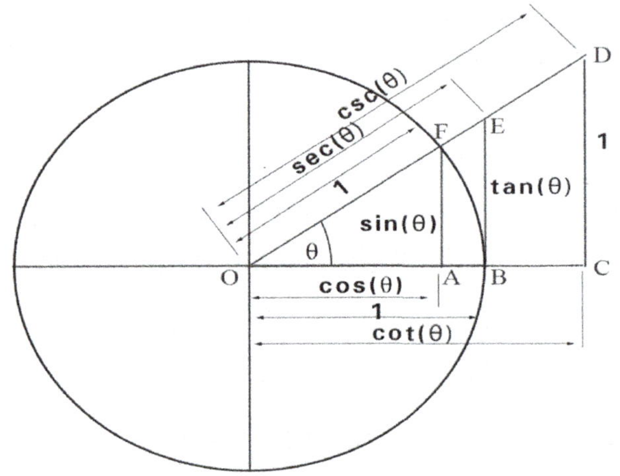

PYTHAGOREAN IDENTITIES

Δ OAF	$\cos^2\theta + \sin^2\theta$	=	1
Δ OBE	$1 + \tan^2\theta$	=	$\sec^2\theta$
Δ OCD	$\cot^2\theta + 1$	=	$\csc^2\theta$

We can now identify the (*x, y*) coordinates of points O, A, B, C, D, E, and F shown in the figure:

O :	(0, 0)	
A :	(cos(θ), 0)	
B :	(1, 0)	
C :	(cot(θ), 0)	$\theta \neq 0, n\pi \;\; n = 1, 2, 3, \ldots$
D :	(cot(θ), 1)	$\theta \neq 0, n\pi \;\; n = 1, 2, 3, \ldots$
E :	(1, tan(θ))	$\theta \neq k\frac{\pi}{2}$ k an odd integer
F :	(cos(θ), sin(θ))	

More About The Unit Circle

Consider a circle of radius 1 centered at (0, 0) on the Cartesian coordinate axis. The *x-y* axis divides the plane into four quadrants; these are numbered with Roman Numerals, counter-clockwise from the positive *x*-axis. As shown to the right, the intersection of the axes with the unit circle define the quadrant angles: "0" on the *x*-axis, 90° on the positive *y*-axis, 180° on the negative *x*-axis, and 270° on the negative *y*-axis. It is important to note that a point (*x, y*) in Quadrant I has both *x* and *y* positive or zero; similarly for Quadrant II, a point

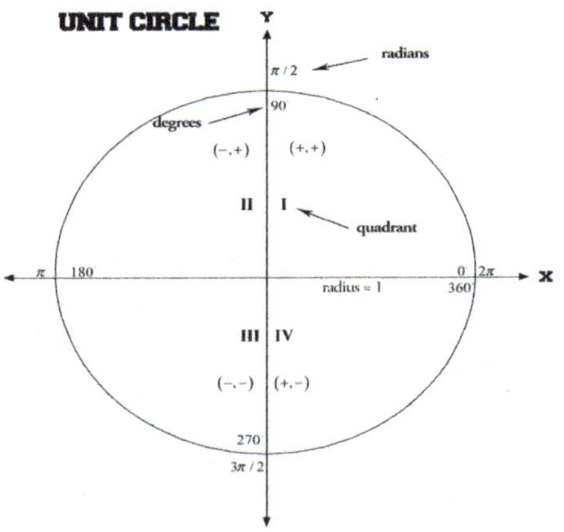

there has x negative and y positive; points in Quadrant III have both x and y negative, and finally, points in Quadrant IV have x positive and y negative. The ordered-pair signs are displayed in the diagram as (+, +), (-, +), (-, -), and (+, -), respectively. The diagram shows the values of the angles both in degrees and in radians. To convert degrees to radians and vice versa, simply apply the following conversion:

$$\pi \text{ radians} = 180 \text{ degrees}.$$

For example, to convert $\pi/2$ radians to degrees, simply divide both sides of the above equation by 2 to get $180/2 = 90$ degrees. Similarly, to convert 30 degrees to radians, divide both sides by 6 (to get $\pi/6$ radians = 30 degrees). That's all there is to it!

Terminal Sides of Angles and Reference Angles

An angle is in *standard position* if its vertex is located at the origin and one ray is on the positive x-axis. The ray on the x-axis is the *initial side* and the other ray is the *terminal side*. The *reference angle* is the acute angle (i.e., between 0 and 90 degrees) formed by the terminal side and the x-axis. Reference angles apply to all quadrants. Angles in quadrant I are their own reference angles (why?); an angle θ whose terminal side falls in Quadrant II has reference angle $\pi - \theta$ (or if in degrees: $180° - \theta$). Angles with terminal sides in Quadrant III have reference angles calculated by $\theta - \pi$ (or $\theta - 180°$). Finally, angles with terminal sides in Quadrant IV have reference angles calculated by $2\pi - \theta$ (*or* $360° - \theta$). All of this is illustrated in the diagram on the previous page.

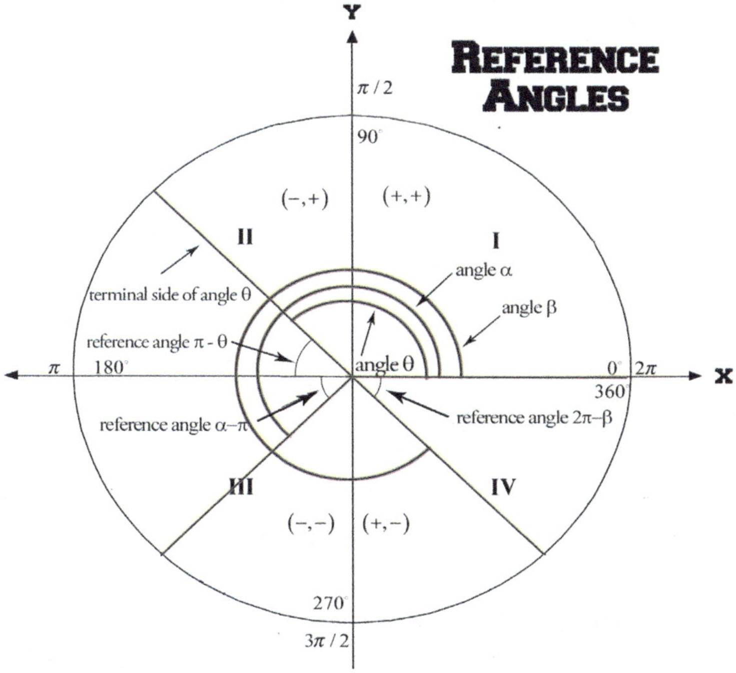

Special angles and the unit circle

Modern trig courses deal frequently with the angles $k\pi/3$, $k\pi/4$, and $k\pi/6$ and the so-called "quadrant angles," $k\pi/2$, where k is an integer. We will use $k = 1$, i.e., positive angles equal-to-or-less than 2π radians (360 degrees); these are listed in the table below. As an exercise in converting between radians and degrees, fill in the blanks: This table seems formidable, but note that if you look at absolute values of the sines and cosines, **there are only three numbers to memorize: 1, 2, and 3.** That means that if you know the signs of x and y for a point in a quadrant, then you need memorize only three numbers to identify the sine and cosine of each of the twelve numbers listed. The numbers for these angles are:

$$\frac{1}{2}; \frac{\sqrt{2}}{2}; \text{ and } \frac{\sqrt{3}}{2}.$$

Quad-rant	Angle		Ref Angle		(sin,cos)	
	Rad	**Deg**	**Rad**	**Deg**	**Sine**	**Co-sine**
I	π/6	30°	π/6	30°	1/2	√3/2
	π/4	45°	π/4	45°	√2/2	
	π/3	60°	π/3	60°		
II	5π/6	150°			1/2	_√3/2
	3π/4	135°			√2/2	_√2/2
		120°	π/3	60°		
III	7π/6	210°	π/6	30°	–1/2	_√3/2
	5π/4	225°	π/4	45°		
		240°	π/3	60°		–1/2
IV	11π/6	330°		30°		
	7π/4	315°	π/4	45°		√2/2
	5π/3	300°	π/3	60°		

To understand where these numbers apply, you need only know that the sine of π/6 (30°) is ½, the sine and cosine of π/4 are equal, and the sine and cosine of π/6 are the cosine and sine of π/3. Verify this from the table above. Let's examine the three Quadrant I standard angles (π/6; π/4; π/3) one at a time in detail. First, π/6. The figure on the next page is self-explanatory. Be sure that you understand every notation in the diagram. Note that the tangent is simply sin/cos. Check it out!

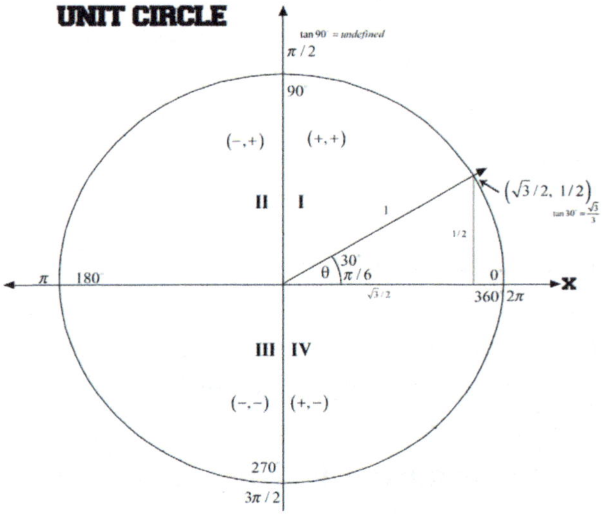

Next we address the standard angle π/4 shown at the top of the next page. Note that the sine and cosine are equal, and that means that the tangent of π/4 is simply 1 (why?). In layman's language referring to the steepness of a road, this is called a "100% grade" - pretty steep!

Last, we examine the standard angle $\pi/3$ below. Note that the sine and cosine of $\pi/3$ are the cosine and sine of $\pi/6$. Now, let's put it all together. The figure on the next page superimposes the three standard angles just discussed. Because we're dealing with reference angles (which are relative to the *x*-axis) we needn't concern ourselves about angles whose terminal sides lie in Quarter II, III, or IV because the numbers are the same - we need only know signs for the quadrant: in Quadrant II, (cos, sin) is (-, +); Quadrant III, (-, -); and Quadramt IV, (+, -).

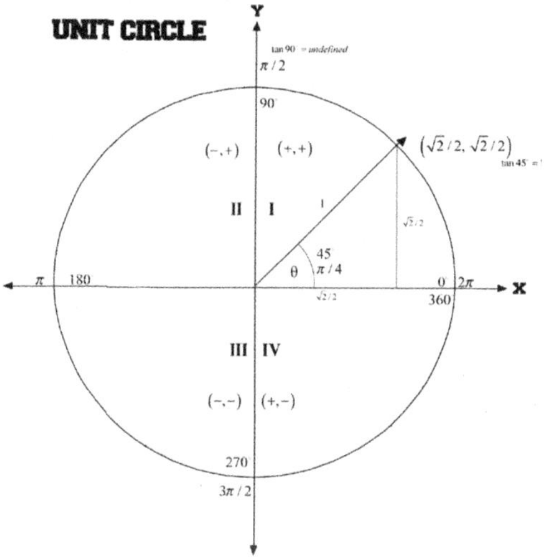

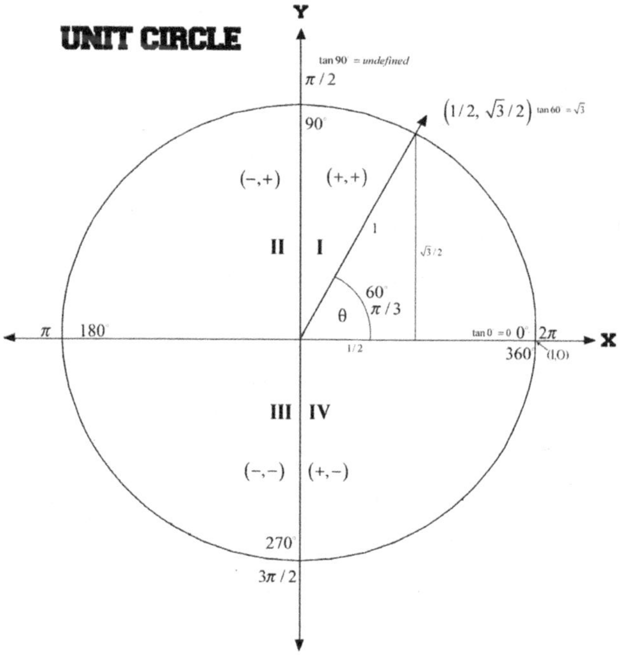

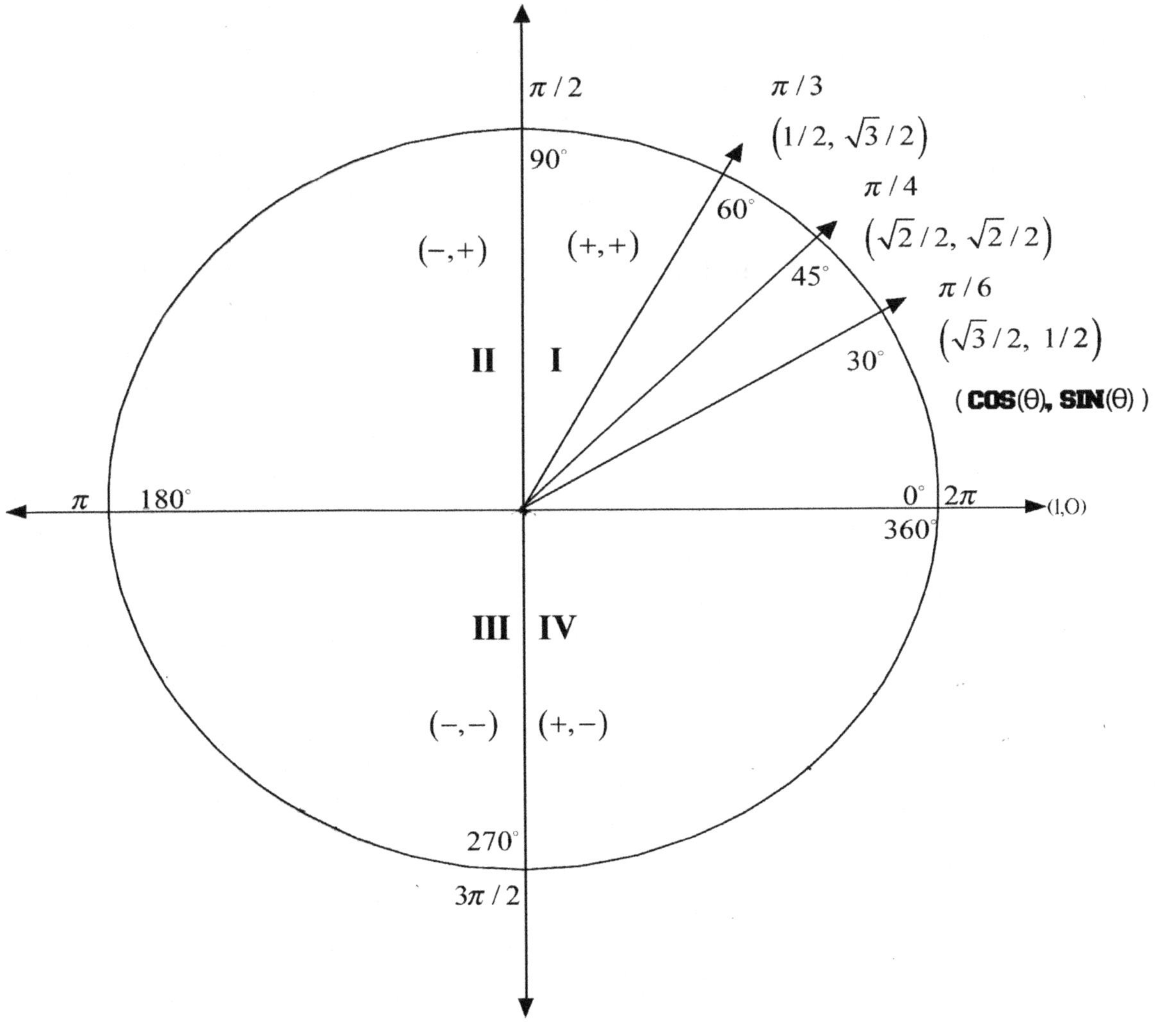
$\pi/2$
90°
$\pi/3$
$(1/2, \sqrt{3}/2)$
60°
$\pi/4$
$(\sqrt{2}/2, \sqrt{2}/2)$
45°
$\pi/6$
$(\sqrt{3}/2, 1/2)$
30°
(**COS**(θ), **SIN**(θ))
(−,+)
(+,+)
II
I
π
180°
0°
2π
360°
(1,O)
III
IV
(−,−)
(+,−)
270°
$3\pi/2$

THE MATH TUTOR'S FRIEND
tangent
i
Trig Identities
It's easy !!
cosine
$e^{i\theta}$
Euler's Formula
De Moivre's formula
Mr. Math
TRIG FUNCTIONS
sine

2. TRIG FUNCTIONS
TABLE OF CONTENTS

●

2. TRIGONOMETRIC FUNCTIONS

Trigonometric ratios (*a review*)

Basic trig ratios and identities*

The trigonometric ratios *sine*, *cosine*, and *tangent*, are defined in terms of the three sides of a *right triangle* as follows (see Figure 2-1):

sine (abbreviated "*sin*") = opposite/hypotenuse;
cosine (abbreviated "*cos*") = adjacent/hypotenuse;
tangent (abbreviated "*tan*") = opposite/adjacent.

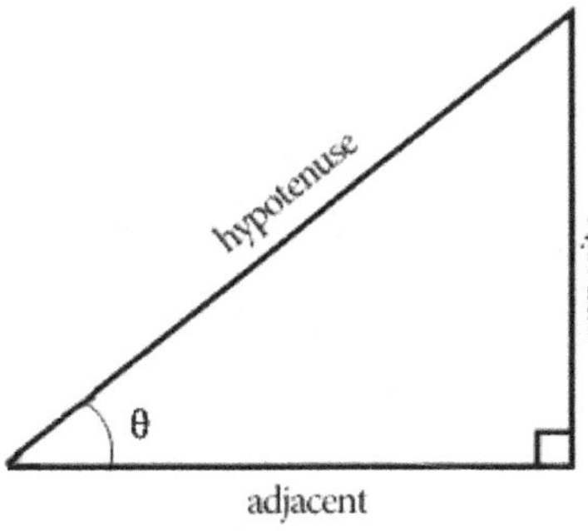

Figure 2-1

Reciprocal trig ratios

sine × *cosecant* = 1;
cosine × *secant* = 1;
tangent × *cotangent* = 1

Here is a mnemonic (memory aid) for the reciprocal trig ratios involving sine and cosine: the *s*s and *c*s go together "**s**ports car"; thus

Sine = 1/***C**osecant*; ***C**osine* = 1/***S**ecant*;.

The reciprocal ratio abbreviations are "*csc*" for "*cosecant* ;" "*sec*" for "*secant* ;" and "*cot*" for "*cotangent* ."

Pythagorean identities

The circle in Figure 2-2 has unit radius. Using the basic trig ratios for sine and cosine defined above, we see that the opposite side is $sin(\theta)$ and the adjacent side is $cos(\theta)$. From the Pythagorean Theorem we have the first Pythagorean identity:

$$sin^2(\theta) + cos^2(\theta) = 1.$$

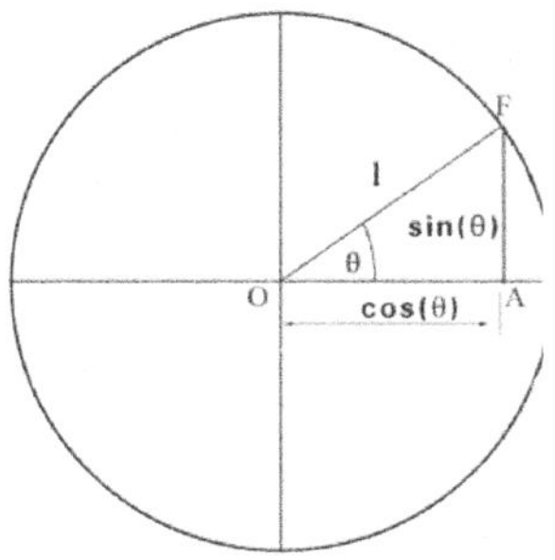

Figure 2-2

Divide this identity by $sin^2(\theta)$ and $cos^2(\theta)$, respectively, to get the remaining two Pythagorean identities:

$$1 + cot^2(\theta) = csc^2(\theta) \quad \text{and} \quad tan^2(\theta) + 1 = sec^2(\theta).$$

Negative angles

Note that the angle θ in Figure 2-3 is measured counterclockwise from the x-axis. Negative angles are measured clockwise. It is clear from the diagram that sin($-\theta$) = -sin(θ), and that cos($-\theta$) = cos(θ). This is a *negative angle identity*.

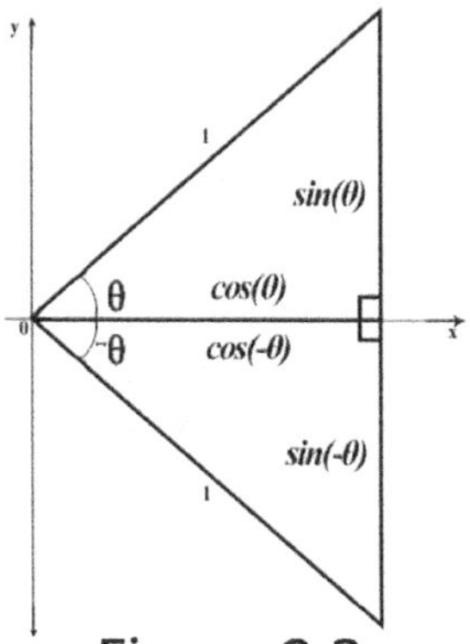

Figure 2-3

* What's the difference between an equation and an identity? An identity is true for any value of the variable, say x or θ, whereas an equation has only one solution for its variable, say x (of course, quadratics have two solutions and so forth, but still only a single specified value in each case). Example: $2x = 4$ is true only for $x = 2$.

Negative angle identities

$$\sin(-\theta) = -\sin(\theta)$$
$$\cos(-\theta) = \cos(\theta)$$

Inverse trigonometric functions

Inverse trigonometric functions are called *arcus functions* or *arc functions* and are usually denoted by *arcsin*(x), *arccos*(x), or *arctan*(x), or more commonly as $sin^{-1}(x)$, $cos^{-1}(x)$, and $tan^{-1}(x)$, respectively. Note that the "-1" superscript does NOT indicate reciprocal. If you want to write the reciprocal of *sin*(x) (which you know is *cosecant*(x)), you would write $sin(x)^{-1}$ or $(sin(x))^{-1}$. Some computer codes denote the arc functions as *asin*, *acos*, and *atan*, respectively.

The expression $y = sin^{-1}(x)$ means "the angle y whose sine is x." You should recognize that if $y = sin^{-1}(x)$, then the value of x must lie between -1 and +1, inclusive, because *sin*(x) [and *cos*(x)] ranges between those values. However, the x in $tan^{-1}(x)$ can take on any value (why?). Here are some equations involving the inverse functions of familiar angles (in radians) within the range of 0 to 2π:

$\pi/4 = \tan^{-1}(1)$ The angle whose tangent is "1" is π/4 (or 5π/4)
$\pi/6 = \sin^{-1}(1/2)$ The angle whose sine is "1/2" is π/6 (or 5π/6)
$\pi/3 = \cos^{-1}(1/2)$ The angle whose cosine is "1/2" is π/3 (or 5π/3)

Law of Sines

The angles and their opposite sides of any triangle (Figure 2-4) are related by the *Law of Sines*:

$$\frac{\sin A}{a} = \frac{\sin B}{b} = \frac{\sin C}{c}$$

Law of Cosines

Referring to the triangle in Figure 2-4, the *Law of Cosines* states:

$$a^2 = b^2 + c^2 - 2bc \times cos(A)$$
$$b^2 = a^2 + c^2 - 2ac \times cos(B)$$
$$c^2 = a^2 + b^2 - 2ab \times cos(C)$$

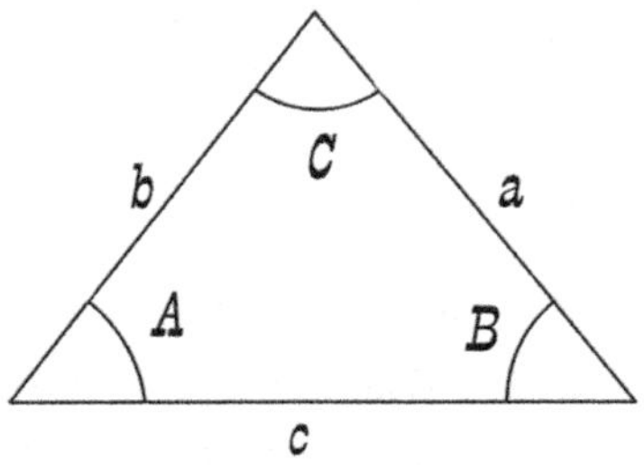

Figure 2-4

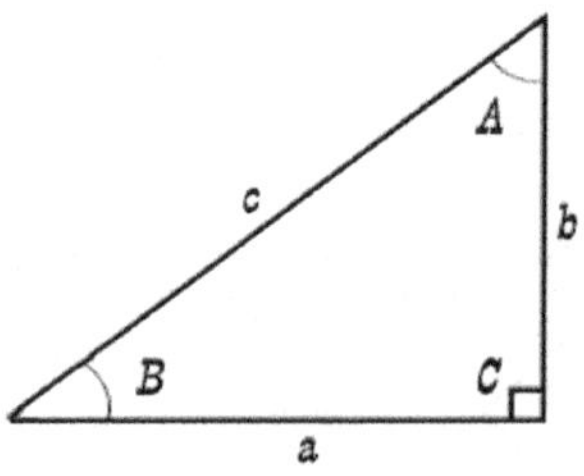

Figure 2-5

Pythagorean Theorem

The *Pythagorean Theorem* is a special case of the Law of Cosines. See Figure 2-5.

For a right triangle with sides a and b and hypotenuse c:

$$c^2 = a^2 + b^2 - 2ab\cos(\pi/2) \quad \text{[Law of Cosines]}$$

$$c^2 = a^2 + b^2 - 2ab\times 0 \quad [\cos(\pi/2) = 0]$$

$$c^2 = a^2 + b^2 \quad \text{[the Pythagorean Theorem.]}$$

Finding angles of a triangle when all sides are known

Refer to Figure 2-4 above. Begin with the Law of Cosines:

$$c^2 = a^2 + b^2 - 2ab\cos(C)$$

Solve for angle C: $2ab\cos(C) = a^2 + b^2 - c^2$

$$\cos(C) = \frac{a^2 + b^2 - c^2}{2ab}$$

$$C = \cos^{-1}\frac{a^2 + b^2 - c^2}{2ab}$$

Where $\cos^{-1}$ is *arccosine* and means "the angle whose cosine is the argument." Once C is known, and since we know the vales of a, b, and c, we can use the Law of Sines to find the remaining angles A and B.

The sine function

Figure 2-6 shows a *sine function*, and we may write $f(t) = \sin(t)$. The sine function is a *periodic function* meaning that it repeats indefinitely. Loosely speaking, the *period* is the the section of the *domain* from a point to the next point where the cycle begins to repeat itself; this is sometimes called one complete *cycle* (in engineering, it is now called a *hertz*). Looking at the graph of the sine function we see that its period is 2π.

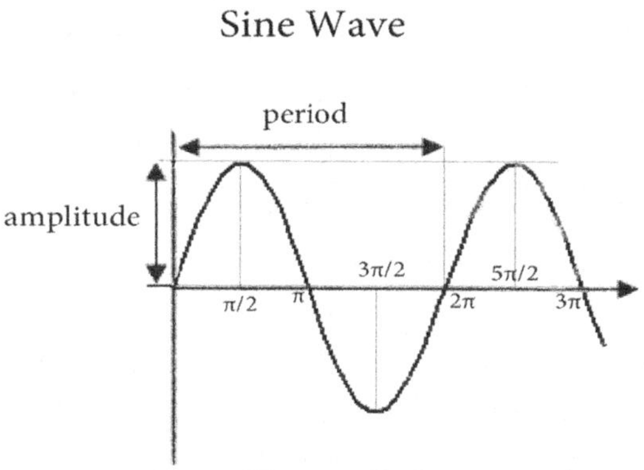

Figure 2-6

That is, the sine function repeats at $2\pi, 4\pi, 6\pi, \ldots k\pi$, where k is an even integer. The *amplitude* of the sine function is 1 unit, measured from the horizontal axis. Note that the *range* of the sine wave is between -1 and 1, inclusive. That is, $-1 \leq \sin(t) \leq 1$.

Figure 2-7 depicts the general sine function $f(x) = a \cdot \sin(bx - c) + d$. As with any function $f(x)$, constants within the parentheses affect horizontal translation and constants outside affect vertical translation; i.e., $f(x - h) + k$ is a horizontal translation to the right of h units and a vertical *translation* of k units. Further, $a \cdot f(bx - h) + k$ is a vertical *stretch* or *shrink* of $|a|$ units and a *dilation* (horizontal stretch or shrink) of $|b|$ units. This information will help in analyzing the sine function. From the preceding discussion we recognize the factors a, b, c, and d

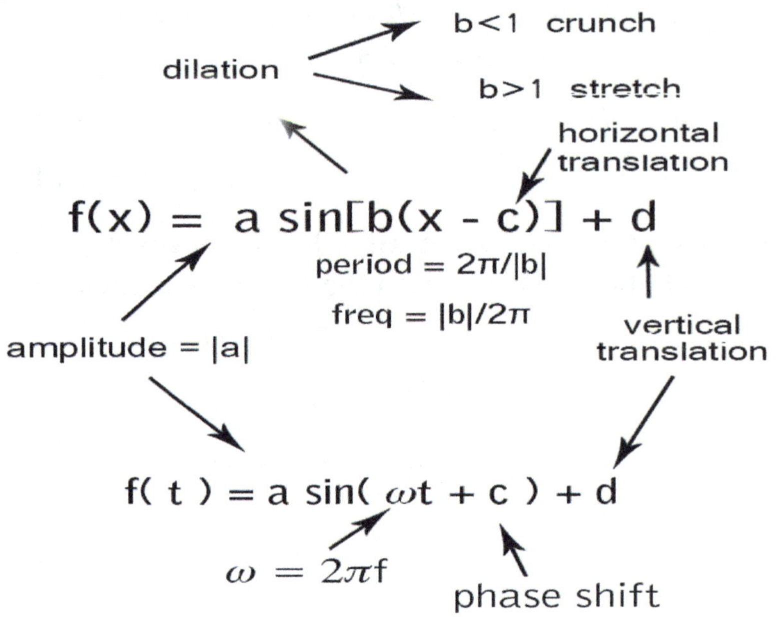

Figure 2-7

as vertical stretch/shrink, dilation, horizontal translation, and vertical translation, respectively. The value $|a|$ is the *amplitude*. Horizontal translation is called *phase shift*. Why do we not address the cosine function? Looking at the graph of the sine function, note that the cosine function is merely the sine function with a phase shift (i.e., horizontal translation) of $\pi/2$ radians.

Complex numbers – a brief review

Let's define the so-called "imaginary number," $\boldsymbol{i} = \sqrt{-1}$ (i.e., $\boldsymbol{i}^2 = -1$). You learned in algebra that the square root of a negative number is undefined, and stated, for example, that there is no solution to quadratics such as $x^2 + 1 = 0$. We define the *complex number* $z = a + b\boldsymbol{i}$, where a and b are *real* numbers and $\boldsymbol{i}$ the *imaginary* number previously discussed. Now, consider a complex plane with the$x$-axis representing real numbers, labeled "*Re*" and the y-axis, labeled "*Im*," representing the imaginary numbers, as shown in Figure 2-8. Locate 3 on the real axis and 2 on the imaginary axis ($\boldsymbol{i}$-axis) and call it 2$\boldsymbol{i}$. Clearly, the point z is at coordinates (3, 2$\boldsymbol{i}$), and if we consider 2, 3$\boldsymbol{i}$, and z to be the distances from the origin of these points, then using *vector addition**, we say that $z = 3 + 2\boldsymbol{i}$, as shown in Figure 2-8.

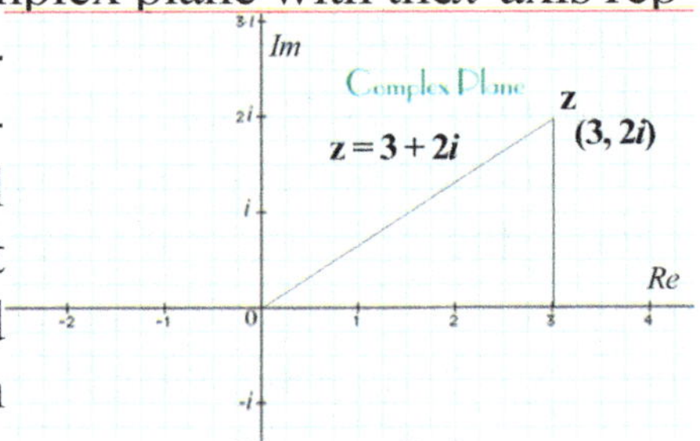

Figure 2-8

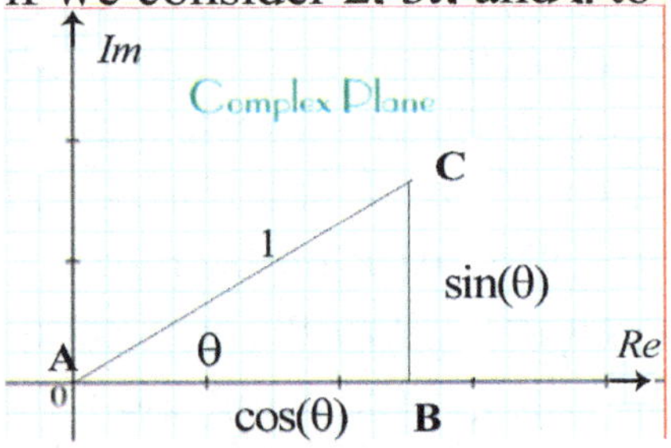

Figure 2-9

Euler's formula

Let's look at the complex plane again, this time appealing to the unit circle (see full discussion in Section 1, "The Unit Circle"); note that the hypot-

enuse of the right triangle ABC in Figure 2-9 is 1, and vertex C is at coordinates ($cos(\theta)$, $i\ sin(\theta)$). Using the definition of z shown above, we see that point z (at C) has coordinates ($cos(\theta)$, $\boldsymbol{i}\ sin(\theta)$). This can be shown by adding the vectors AB and BC to yield vector AC. Point C can be expressed as $\boldsymbol{e}^{i\theta}$; that is, the point on the unit circle having angle of θ radians from the positive *Re*-axis.
Using the definitions of sine and cosine, this time recognizing that the sine is the distance on the imaginary axis, we can now introduce *Euler's formula*:

$$\boldsymbol{e}^{i\theta} = cos(\theta) + \boldsymbol{i}\ sin(\theta)$$

abbreviated as "$cis(\theta)$." An explanation of Euler's Formula involves *power series*, a topic beyond the scope of this discussion.

Sum of angles identities
Let's look at the sum of angles in order to derive a new set of trig identities. Figure 2-10 shows two angles, α and θ (alpha and theta).

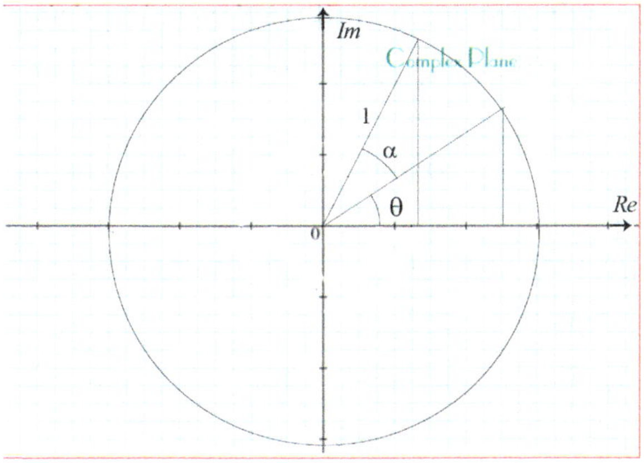

Figure 2-10

From a prior discussion we know that

$$\boldsymbol{e}^{i\theta} = cos(\theta) + i\ sin(\theta),$$

and

$$\boldsymbol{e}^{i\alpha} = cos(\alpha) + i\ sin(\alpha).$$

Taking the sum of α and θ, we use Euler's formula and the laws of exponents:

$$\boldsymbol{e}^{i(\alpha+\theta)} = \boldsymbol{e}^{i\alpha}\,\boldsymbol{e}^{i\theta} = (cos(\alpha) + i\ sin(\alpha))(\ cos(\theta) + i\ sin(\theta))$$

Multiply the binomials using FOIL:

$$\boldsymbol{e}^{i(\alpha+\theta)} = cos(\alpha)cos(\theta) + cos(\alpha)i\ sin(\theta) + i\ sin(\alpha)cos(\theta) + i\ sin(\alpha)\ i\ sin(\theta)$$

$$= cos(\alpha)\ cos(\theta) + \boldsymbol{i}^2 sin(\alpha)sin(\theta) + i\ [cos(\alpha)\ sin(\theta) + sin(\alpha)cos(\theta)].$$

Recognizing that $\boldsymbol{i}^2 = -1$, the above equation becomes

$$\boldsymbol{e}^{i(\alpha+\theta)} = cos(\alpha)\ cos(\theta) - sin(\alpha)\ sin(\theta) + i\ [cos(\alpha)\ sin(\theta) + sin(\alpha)cos(\theta)].$$

But from Euler's formula,

$$e^{i(\alpha+\theta)} = cos(\alpha + \theta) + i\,sin(\alpha + \theta),$$

so that

$$cos(\alpha+\theta) + i\,sin(\alpha+\theta) = cos(\alpha)cos(\theta) - sin(\alpha)sin(\theta) + i\,[cos(\alpha)sin(\theta) + sin(\alpha)cos(\theta)]$$

Now, the real part of the left side of the equation must equal the real part of the right side; and similarly for the imaginary parts, therefore,

$$cos(\alpha + \theta) = cos(\alpha)cos(\theta) - sin(\alpha)sin(\theta)$$

and

$$sin(\alpha + \theta) = cos(\alpha)sin(\theta) + sin(\alpha)cos(\theta).$$

These are the *sum of angles identities*. How about difference of angles? Let's apply the negative angle identities already discussed: we know that

$sin(-\theta) = -sin(\theta)$, and that $cos(-\theta) = cos(\theta)$.

Applying the negative angle identities to the sum-of-angles expressions, we have

$$cos(\alpha - \theta) = cos(\alpha)cos(-\theta) - sin(\alpha)sin(-\theta),$$

which becomes

$$cos(\alpha - \theta) = cos(\alpha)cos(\theta) + sin(\alpha)sin(\theta)$$

and

$$sin(\alpha - \theta) = cos(\alpha)sin(-\theta) + sin(\alpha)cos(-\theta),$$

which, after rearranging terms. becomes

$$sin(\alpha - \theta) = sin(\alpha)cos(\theta) - cos(\alpha)sin(\theta).$$

Sum of angles identities

$$cos(\alpha + \theta) = cos(\alpha)cos(\theta) - sin(\alpha)sin(\theta)$$
$$sin(\alpha + \theta) = cos(\theta)sin(\alpha) + sin(\alpha)cos(\theta)$$
$$cos(\alpha - \theta) = cos(\alpha)cos(\theta) + sin(\alpha)sin(\theta)$$
$$sin(\alpha - \theta) = sin(\alpha)cos(\theta) - cos(\alpha)sin(\theta).$$

De Moivre's formula

Here's a neat way, known as DeMoivre's formula, to avoid binary expansion and express the power of $cis(\theta)$ to a cosine and sine:

$$cis(\theta)^n = [cos(\theta) + i\,sin(\theta)]^n = cos(n\theta) + i\,sin(n\theta)$$

The above identity holds only when "n" is an integer.
The formula can also be written as

$$cis(\theta)^n = cis(n\theta).$$

Arc length of a circle

The (linear) length, s. of an arc of a circle with radius r intercepted by a central angle θ (measured in radians) is given by the formula

$$s = r\theta$$

[This is easily verified by setting the angle to 2π to get $s = C$, the circumference of the circle.]

Tips for simplifying trigonometric identities

Work the "hard" side first to try to simplify to the "easy" side. If no progress, try converting everything to sines and cosines.

Examples

(1) Show that $sin^4(x) - cos^4(x) = (sin(x) - \cos(x))(sin(x) + cos(x))$.

Solution: We recognize that the "hard" side, on the left, is the difference of two squares in $sin^2(x)$:

$$sin^4(x) - cos^4(x) = (sin^2(x) - cos^2(x))(sin^2(x) + cos^2(x))$$

The term $(sin^2(x) + cos^2(x))$ is simply the familiar Pythagorean identity, and equals 1; the term $(sin^2(x) - cos^2(x))$ is also a difference of two squares, so

$$sin^4(x) - cos^4(x) = (sin^2(x) - cos^2(x))(1)$$
$$= (sin(x) - \cos(x))(sin(x) + cos(x)).$$

Note that we have proved also that

$$sin^4(x) - cos^4(x) = sin^2(x) - cos^2(x).$$

Check it out on your calculator!

(2) Write $cos(x)$ in terms of $sin(x)$.

Solution: From the Pythagorean identity we know that

$$sin^2(x) + cos^2(x) = 1$$

So, just solve the above for $\cos(x)$ to get

$$cos(x) = \pm\sqrt{(1 - sin^2(x))}$$

(3) Show that

$$cos^2(x)/(1 + sin(x)) = 1 - sin(x)$$

Solution: This is a *rational function*. Before attempting a solution we must determine its domain. Since the denominator cannot be zero, then $sin(x)$ cannot have value of -1, and this would occur if x had value $3\pi/2$ (restricting x to the interval $[0, 2\pi)$. Now apply the Pythagorean identity to the numerator and you can cancel the $1+ sin(x)$.

(4) Show that

$$\sin^2(x)\,\tan^2(x) = \tan^2(x) - \sin^2(x)$$

Solution: There are several ways to approach this; I simply change the tangents to sines and cosines to see that the result is one of the Pythagorean identities. Try it!

EXERCISES

Prove the following identities:

a) $\cos(x)(1 + \sec(x)) = 1 + \cos(x)$.

b) $(1 - \cos^2(x))(1 + \cot^2(x)) = 1$

c) *[difficult]* Expand the following expression:

$$(\cos(x) + \boldsymbol{i}\sin(x))^3 + (\cos(\theta) + \boldsymbol{i}\sin(\theta))^3$$

(HINT: Recognize the sum of cubes, apply F.O.I.L., De Moivre's formula, and then collect terms)

●

Summary of identities & formulas

Reciprocal trig ratios

$$sine \times cosecant = 1;$$

$$cosine \times secant = 1;$$

$$tangent \times cotangent = 1$$

Pythagorean identities

$$sin^2(\theta) + cos^2(\theta) = 1.$$

$$1 + cot^2(\theta) = csc^2(\theta).$$

$$tan^2(\theta) + 1 = sec^2(\theta).$$

Euler's formula

$$e^{i\theta} = cos(\theta) + i\ sin(\theta).$$

Sum of angles identities

$$cos(\alpha + \theta) = cos(\alpha)cos(\theta) - sin(\alpha)sin(\theta)$$

$$sin(\alpha + \theta) = cos(\theta)sin(\alpha) + sin(\alpha)cos(\theta)$$

$$cos(\alpha - \theta) = cos(\alpha)cos(\theta) + sin(\alpha)sin(\theta)$$

$$sin(\alpha - \theta) = sin(\alpha)cos(\theta) - cos(\alpha)sin(\theta)$$

De Moivre's formula

$$(cos(\theta) + \boldsymbol{i}sin(\theta))^n = cos(n\theta) + \boldsymbol{i}sin(n\theta)$$

or equivalently

$$(cis(\theta))^n = cis(n\theta).$$

where n is an integer.

•

THE MATH TUTOR'S FRIEND
vector
displacement
vector length
It's easy !!
unit vector
POLAR COORDINATES & VECTORS
scalar
||r||
Mr. Math

3. POLAR COORDINATES & VECTORS

TABLE OF CONTENTS

•

3. POLAR COORDINATES AND VECTORS

Polar coordinates

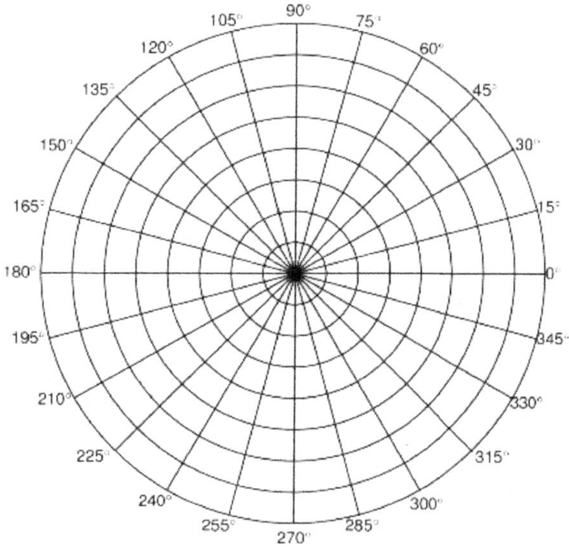

In addition to the *x-y* Cartesian coordinate system (and other systems) there is the *polar coordinate system*, illustrated above, in which direction is indicated by angles. We shall denote the radial coordinate by ***r***, and the angular coordinate by θ (Greek letter *theta*). Angles in polar notation may be expressed in *degrees*, *radians* (2π radians equal 360°), or *gradians* (*grads*) (400 grads in a circle). Degrees are traditionally used in navigation, surveying, and many applied disciplines, while radians are more common in mathematics and physics; grads are used in some military and surveying applications and will not be addressed further here.

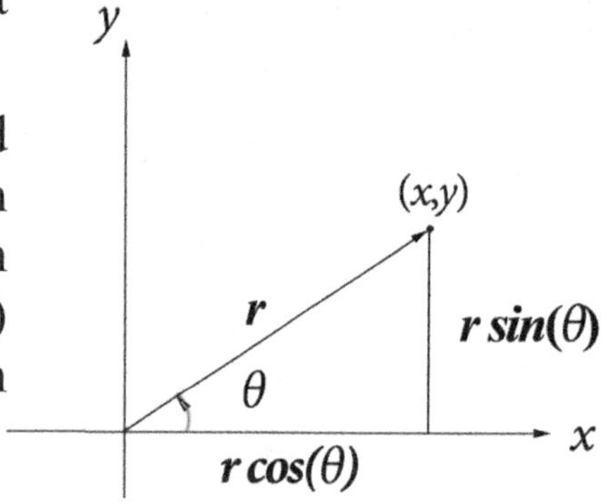

In the polar coordinate system the angle θ is measured counterclockwise from the axis or reference side. In mathematical literature, the *polar axis* is often drawn horizontal and points to the right*. The diagram (right) illustrates the relationship between polar and Cartesian coordinates.

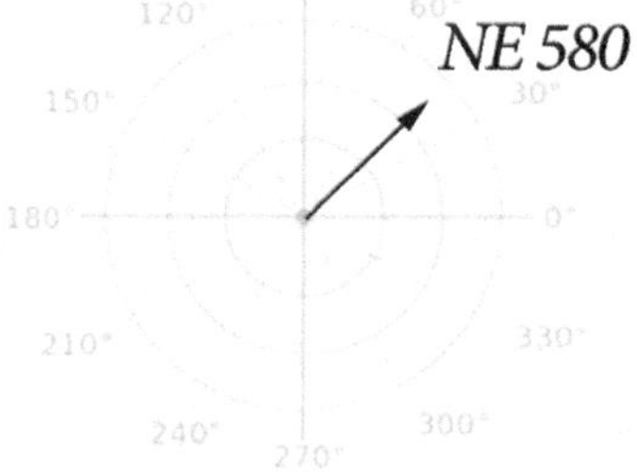

Vectors

A *vector* has magnitude and direction; for example, an airplane traveling at 580 mph heading northeast. A vector is an arrow representing magnitude (i.e., speed) and direction; its length shows the amount and the arrowhead shows direction. Speed does not

* In some military applications, the navy, for example, the polar axis is vertical to represent true north; in other cases the polar axis represents the heading of a ship or aircraft as reference.

imply direction, thus speed is a *scalar*. A vector is indicated by an arrow with a defined length; its length shows the amount and the arrowhead shows direction.

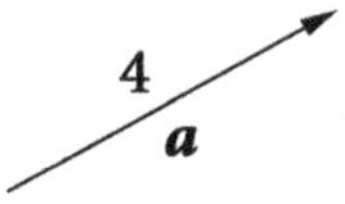

Representing vectors

A vector from point A to point B may be denoted by the notation $(r, \angle\theta)$ or simply (r, θ) where r is the length of the vector and θ is its angle. We can write its length as AB without the bar or $\|\overline{\mathbf{AB}}\|$

In our review of matrices we saw that each row of a matrix is a vector, and each column of a matrix is a vector. For an m×n matrix **A**, any column j, $1 \le j \le$ n may be represented by the vector $\mathbf{a}_j = (a_{j1}, a_{j2}, \ldots a_{jn})$ and any row i, $1 \le i \le$ m may be represented by the vector $\mathbf{a}_i = (a_{i1}, a_{i2}, \ldots a_{im})$. Other notation for a vector: $< 2, 3 >$ is $2\hat{\imath} + 3\hat{\jmath}$ is $2\hat{x} + 3\hat{y}$

Displacement *vs* distance

We saw that vector addition produces the *resultant*, shown in the figure to the right as the vector sum of *displacements*. For example, the "as the crow flies" distance between two towns is 20 miles. Departing one town for the other one might take the "scenic route" which meanders over back roads that eventually lead to the destination after, say, a trip of 37 miles. The distance, 37, is a scalar and the "as the crow flies" is a vector of length 20.

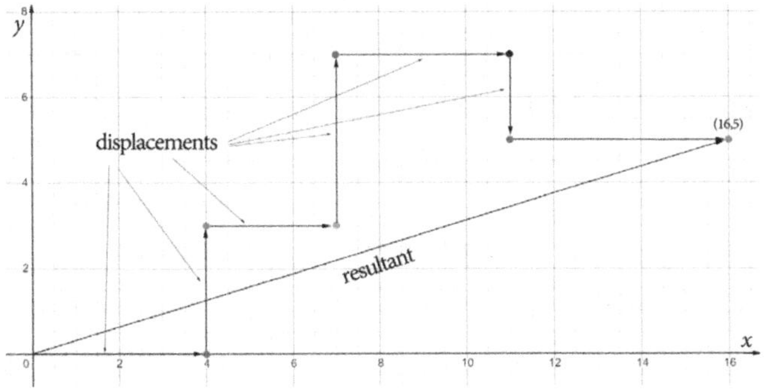

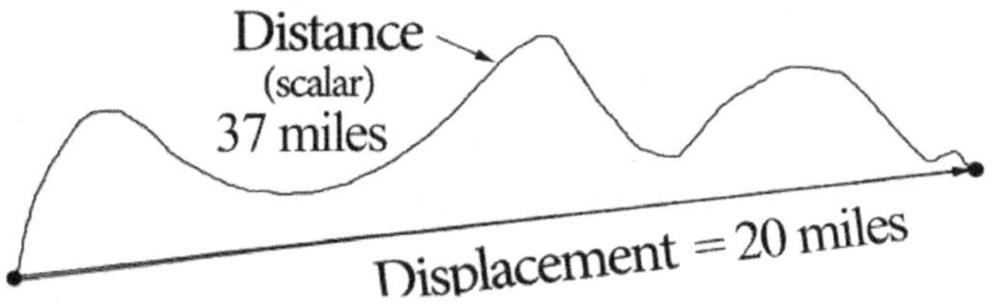

Vector notation

A scalar is just a number - a quantity like temperature or distance. It's a magnitude with no indication of direction. In this book scalar variables will be in lower case or lower case italics, sometimes with subscripts, e.g., a, a_2, u_1, etc. A vector, which indicates magnitude and direction can be boldface (e.g., **u**, **V**, **A**), or letters capped with "hats" or arrows (e.g., ŷ, $\vec{A}$, $\underline{\mathrm{A}}$); we shall generally use boldface letters to represent vectors, e.g., **u**.

Vector length (magnitude)

Denote the magnitude of vector **u** by ||**u**||. For **u** = $<u_1, u_2>$, $\|u\| = \sqrt{u_1{}^2 + u_2{}^2}$

Note: the magnitude of a vector, say **u**, may be indicated by |**u**| or ||**u**||. The latter is preferable to avoid confusion with the absolute-value symbol, e.g., |u|.

Unit vector

If **a** is a vector, we write **â** ("*a* hat") to represent a *unit vector* in the direction of **a**. Vectors may be in 2-, 3-, or n-dimension space. In mathematics the concept of dimensions greater than three is easily handled; dimensions are simply whole numbers after all. Indeed, n-dimensional geometry does exist. Here, however, we limit ourselves to 2- and 3-dimensions. Vectors in these dimensions may be expressed as **v** = < v_1, v_2 >, a vector whose endpoints are the origin (0, 0) and point (v_1, v_2). Ditto for the three-dimension vector **w** = < w_1, w_2, w_3 >. Every nonzero vector has a corresponding *unit vector* in the same direction with magnitude 1. To find the unit vector û of the vector divide that vector by its magnitude:

$$\hat{u} = \mathbf{u}/\|\mathbf{u}\|.$$

For two dimensions, denote unit vectors in the x- and y-directions by $\hat{\imath}$ and $\hat{\jmath}$,

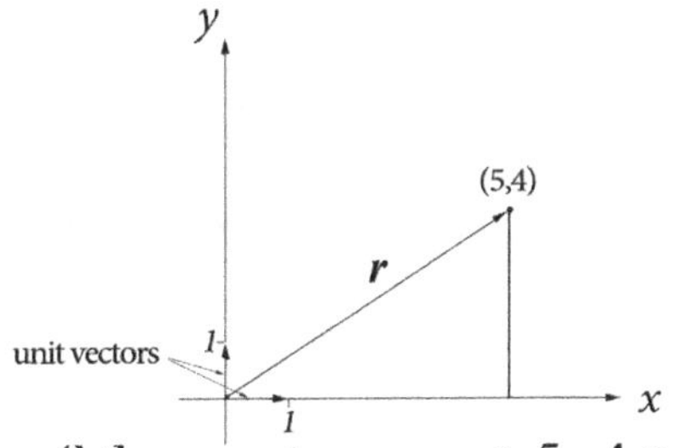

respectively. The length of the vector r = < 5, 4 > = 5**x** + 4**y** is given by the distance formula:

$$\|r\| = \sqrt{5^2 + 4^2} = \sqrt{41} \approx 6.4.$$

$$unit\ vector = \frac{vector}{vector\ length}$$

The vector $\mathbf{u} = \langle 3,4 \rangle$ has unit vector $\hat{\mathbf{u}} = \frac{\mathbf{u}}{\|\mathbf{u}\|}$

$$= \langle \frac{3}{\sqrt{3^2 + 4^2}}, \frac{4}{\sqrt{3 + 4^2}} \rangle = \langle 3/5, 4/5 \rangle.$$

Adding vectors

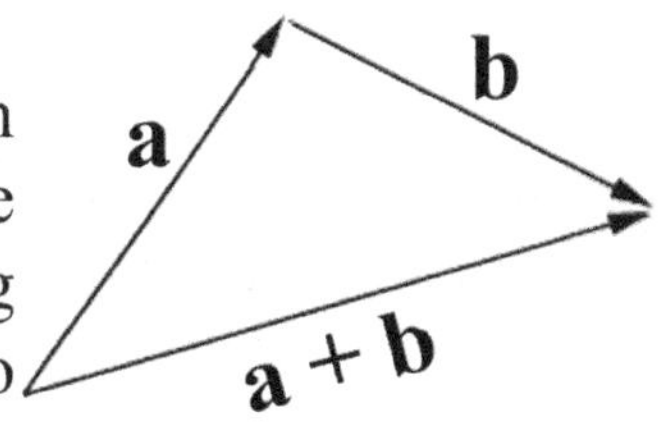

One of the things we can do with vectors is to add them together. Informally, adding vectors is simply to translate them end to end where the end of one vector is the starting point for the next ("head-to-tail"). We start by adding two vectors, say **a** and **b**, together. The sum of the vectors, **a** + **b** (or the *resultant*, as it is sometimes called) is what we get when we complete the triangle by constructing the resultant vector from the origin of the first vector to the tail of the last, as shown in the figure above. This is the *triangle law for adding vectors*. We can add any number of vectors together by adding the first two, then adding the result to the third, and so on. There is another way of adding two vectors, as shown to the right. Instead of making the second vector start where the first one finishes, use a *common origin* and construct a parallelogram. The diagonal of the parallelogram is the sum **a** + **b**. This is the *parallelogram law for adding vectors*.

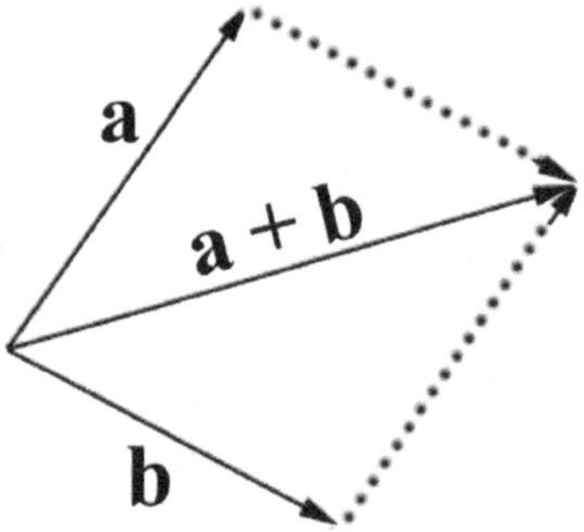

Unit vectors may be used to represent the axes of a Cartesian coordinate system. For instance, the standard unit vectors in the direction of the *i*, *j*, and *k* axes of a three dimensional Cartesian coordinate system are

$$\hat{\mathbf{i}} = \begin{bmatrix} 1 \\ 0 \\ 0 \end{bmatrix}, \ \hat{\mathbf{j}} = \begin{bmatrix} 0 \\ 1 \\ 0 \end{bmatrix}, \ \hat{\mathbf{k}} = \begin{bmatrix} 0 \\ 0 \\ 1 \end{bmatrix}$$

$$\hat{\imath} = (1, 0, 0); \ \hat{\jmath} = (0, 1, 0); \text{ and } \hat{k} = (0, 0, 1)$$

These form a set of mutually orthogonal unit vectors. That is, each is at right angles to the other two; this, of course, can be only in three dimensions. Any two unit vectors fall in a plane. A unit vector has length 1, so that $\|\hat{a}\| = 1$ (sometimes written $|\hat{a}| = 1$).

Right-hand rule. This rule states that a right hand closing $\hat{\imath}$ into $\hat{\jmath}$ will have the thumb extending into the $\hat{k}$ direction.

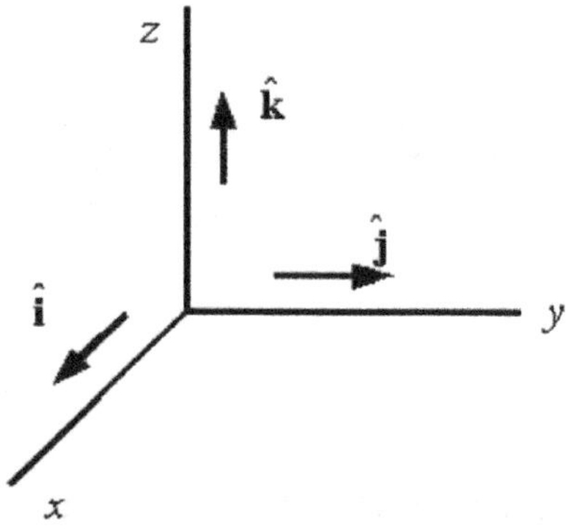

Representing a vector in terms of a unit vector
The diagram below shows vector **A** with length A in the direction of the unit vector ***e***. That is,

$$\mathbf{A} = A\boldsymbol{e}$$

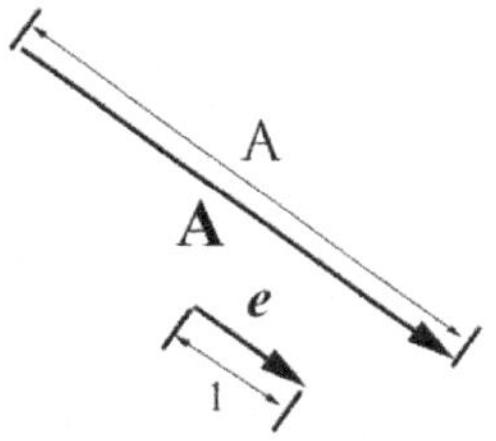

[Note that ***e***, as used here, is merely a label (of a unit vector) and is not related to the Euler constant ***e***]

Base vectors

Base vectors are sets of unit vectors that represent the directions of corresponding

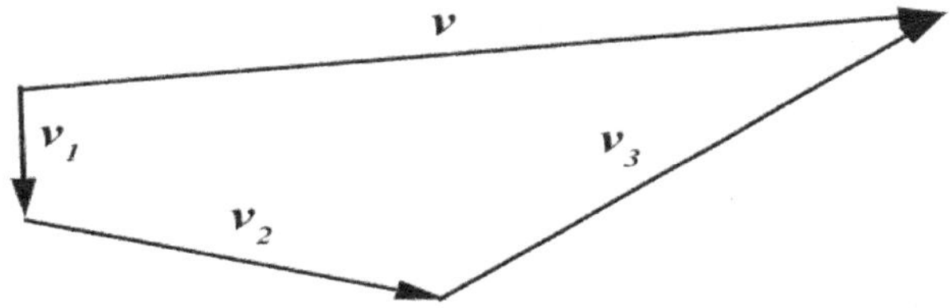

sets of (non-unit) vectors. The diagram shown above shows vector $\boldsymbol{v}$ as the *resultant* of the three vectors $\boldsymbol{v}_1$, $\boldsymbol{v}_2$, and $\boldsymbol{v}_3$; that is

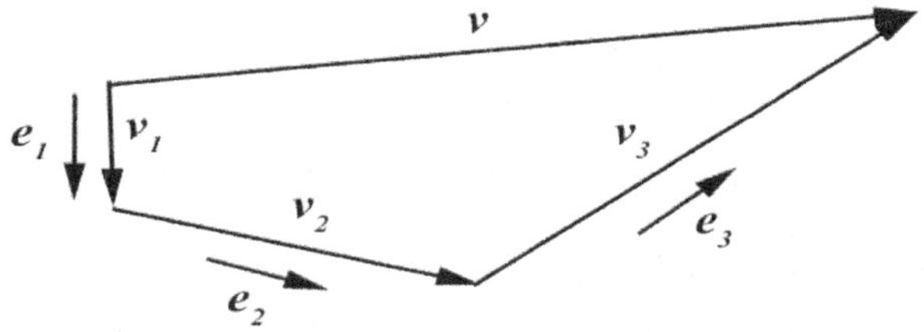

$$\boldsymbol{v} = \boldsymbol{v}_1 + \boldsymbol{v}_2 + \boldsymbol{v}_3$$

each of which has associated unit vectors $\boldsymbol{e}_1$, $\boldsymbol{e}_2$, and $\boldsymbol{e}_3$ as shown below.

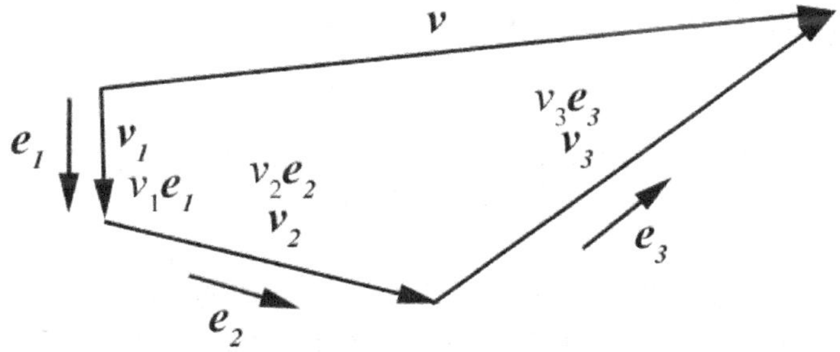

The magnitudes of $\boldsymbol{v}_1$, $\boldsymbol{v}_2$, and $\boldsymbol{v}_3$ are v_1, v_2, *and* v_3, and so we have

$$\boldsymbol{v}_1 = v_1\boldsymbol{e}_1 \text{ and } \boldsymbol{v}_2 = v_2\boldsymbol{e}_2 \text{ and } \boldsymbol{v}_3 = v_3\boldsymbol{e}_3.$$

The original (resultant) vector can be expressed as follows:

$$\boldsymbol{v} = \boldsymbol{v}_1 + \boldsymbol{v}_2 + \boldsymbol{v}_3 = v_1\boldsymbol{e}_1 + v_2\boldsymbol{e}_2 + v_3\boldsymbol{e}_3$$

The coefficients of the unit vectors are scalar multipliers which denote magnitudes of the component vectors v_1, v_2, and v_3 resulting in v.

Scalar multiplication

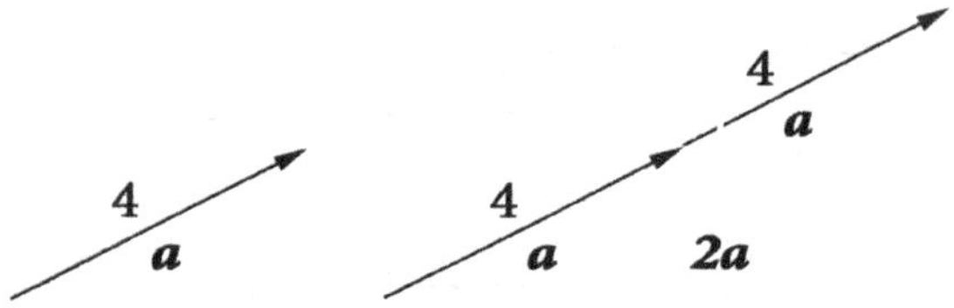

As described above, multiplication of a vector $\boldsymbol{a}$ by a *scalar multiplier* (or *scalar*) changes (i.e., “scales”) the length of that vector and does not affect its direction.

Angle between vectors

Suppose that **u** and **v** are two vectors having a common origin, i.e., placed so that their initial points are the same. Then the angle θ between **u** and **v** such that $0 < \theta < \pi$. See the examples shown below.

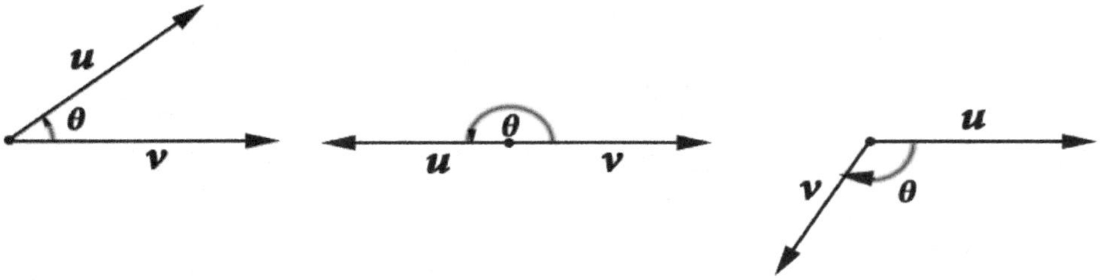

Notice that there are two angles formed by the two vectors and the one that we will always choose is the one that satisfies $0 \leq \theta \leq \pi$.

Multiplying vectors - dot product

Let **u** and **v** be two vectors in 2-space and θ the angle between them (we'll dispense with the "hat" notation):

$$\mathbf{u} = a\mathbf{i} + b\mathbf{j} \text{ and } \mathbf{v} = c\mathbf{i} + d\mathbf{j}$$

The *dot product* of the two vectors is given by

$$\mathbf{u} \cdot \mathbf{v} = ac + bd$$

Example: Find the dot product of $\mathbf{v} = 3\mathbf{i} + 5\mathbf{j}$ and $\mathbf{w} = 2\mathbf{i} + 3\mathbf{j}$

$$\mathbf{v} \cdot \mathbf{w} = (3)(2) + (5)(3) = 21$$

The dot product is called a *scalar product* because it produces a scalar. Two vectors **u** and **v** are *orthogonal* if their angle is a right angle; that is $\mathbf{u} \cdot \mathbf{v} = 0$ (why?)

Relationship of dot product to angle between vectors

The relation of the angle, θ, and the two vectors, **u** and **v**, is given by

$$\cos\theta = \frac{u \cdot v}{||u||\ ||v||}$$

so that the dot product is defined by

$$\mathbf{u} \cdot \mathbf{v} = ||\mathbf{u}||\ ||\mathbf{v}|| \cos\theta.$$

Example: Find the angle between $\mathbf{u} = 2\mathbf{i} + 3\mathbf{j}$ and $\mathbf{v} = 4\mathbf{i} + \mathbf{j}$

The dot product: $\mathbf{u} \cdot \mathbf{v} = (2)(4) + (3)(1) = 11$

and

$$||u|| = \sqrt{2^2 + 3^2} = \sqrt{13}$$

$$||v|| = \sqrt{4^2 + 1^2} = \sqrt{17}$$

$\theta = arccosine(11/\sqrt{13}\sqrt{17}) = 42°$, approximately.

Vector projections

Let **u** and **v** be a vectors. Then **u** can be broken up into two components, **r** and **s** such that **r** is parallel to **v** and **s** is perpendicular to **v**. The component **r** = proj_v**u** is the *projection* of **u** onto **v** and **s** is the component of **u** perpendicular to **v**.

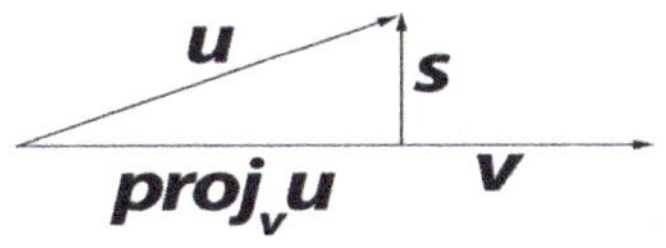

$$\mathbf{u}\cdot\mathbf{v} = \|\mathbf{u}\|\,\|\mathbf{v}\|\cos\theta = \frac{\|\mathbf{u}\|\,\|\mathbf{v}\|\,\|\text{proj}_v\mathbf{u}\|}{\|\mathbf{u}\|}$$

$$= \|\mathbf{v}\|\,\|\text{proj}_v\mathbf{u}\|$$

We can calculate the projection of **u** onto **v** by the formula $\|\text{proj}_v\mathbf{u}\| = \mathbf{u}\cdot\mathbf{v}/\|\mathbf{v}\|$. The direction is correct since the right hand side of the formula is a constant multiple of **v** so the projection vector is in the direction of **v** as required. To find the vector s, notice from the diagram that the vector addition is $\text{proj}_v\mathbf{u} + \mathbf{s} = \mathbf{u}$, therefore,

$$\mathbf{s} = \mathbf{u} - \text{proj}_v\mathbf{u}.$$

Example: When pushing a box on a horizontal floor, say, with the stick at an angle θ with the floor and ignoring friction and gravity, how much of the effort (***v***) contributes to moving the box along the floor (vector ***u***), and how much is wasted "pushing down" against the floor? That is, how much of one vector (the force on the stick) projects in the horizontal direction along the floor?

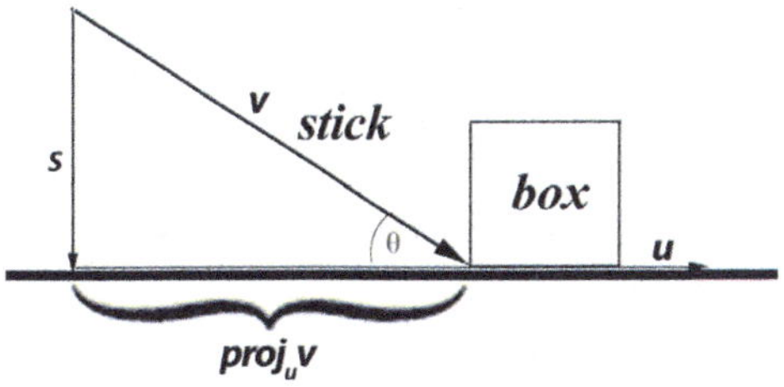

The stick is the vector projection of ***v*** onto ***u***, proj_uv, the direction of **u** (horizontal). Vector ***v*** resolves into two component vectors, one that is parallel to the the floor (***u***) and ***s***, perpendicular to ***u***. The parallel vector, ***u***, is ***proj$_u$v***. Pushing the box at an angle $\theta > 0$ results in wasted effort in the amount $s = |v|\sin\theta$, the remaining energy doing the intended work. As with all vectors, ***u*** and ***v*** have both magnitude and direction.

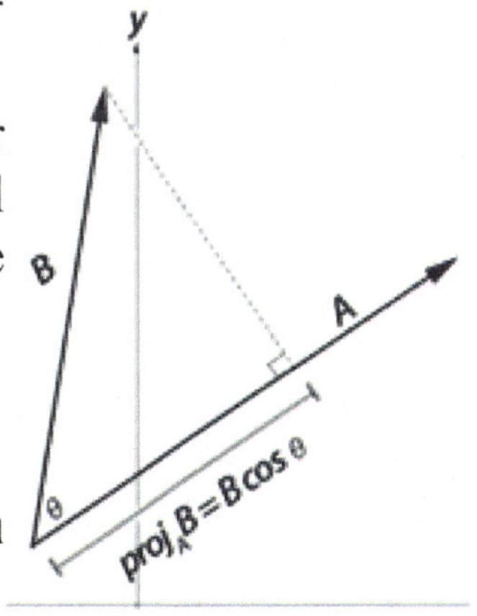

Note that although this example deals with the vector projection on the x-axis, a vector may have an orthogonal projection on any other vector, as shown to the right. The magnitude of a projection is the *scalar projection.*

Properties of the dot product

1. $\mathbf{u}\cdot\mathbf{v} = \mathbf{v}\cdot\mathbf{u} = \|\mathbf{u}\|\,\|\mathbf{v}\|\cos\theta$ (commutative)
2. If $\mathbf{u}\cdot\mathbf{v} = 0$ then **u** or **v** (or both) are zero or **u** and **v** are orthogonal ($\cos 90° = 0$)

3. $(a\mathbf{u})\cdot(b\mathbf{v}) = (b\mathbf{u})\cdot(a\mathbf{v}) = ab\mathbf{u}\cdot\mathbf{v}$ (here, a and b are scalars)
4. $\mathbf{u}\cdot\mathbf{u} = ||\mathbf{u}||^2$ (this means that $\boldsymbol{i}^2 = \boldsymbol{j}^2 = \boldsymbol{k}^2 = 1$)
5. $\mathbf{u}\cdot(\mathbf{v} + \mathbf{w}) = \mathbf{u}\cdot\mathbf{v} + \mathbf{u}\cdot\mathbf{w}$
6. If $\mathbf{u} = \langle u_1, u_2, u_3\rangle$ and $\mathbf{v} = \langle v_1, v_2, v_3\rangle$ then $\mathbf{u}\cdot\mathbf{v} = u_1v_1 + u_2v_2 + u_3v_3$

Direction cosines

Let $\mathbf{v} = a\mathbf{i} + b\mathbf{j} + c\mathbf{k}$ be a vector, the *direction cosines* of $\mathbf{v}$ are its cosines along the three coordinate axes:

$$\cos(\alpha) = a/||\mathbf{v}||$$
$$\cos(\beta) = b/||\mathbf{v}||$$
$$\cos(\gamma) = c/||\mathbf{v}||$$

Multiplying vectors

cross product

Let $\mathbf{u} = a\mathbf{i} + b\mathbf{j} + c\mathbf{k}$ and $\mathbf{v} = d\mathbf{i} + e\mathbf{j} + f\mathbf{k}$ be vectors; we define the *cross product* $\mathbf{u} \times \mathbf{v}$ by the determinant of the matrix:

$$\begin{pmatrix} i & j & k \\ a & b & c \\ d & e & f \end{pmatrix}$$

We can compute this determinant using cofactors as

$$\begin{vmatrix} b & c \\ e & f \end{vmatrix}\boldsymbol{i} - \begin{vmatrix} a & c \\ d & f \end{vmatrix}\boldsymbol{j} + \begin{vmatrix} a & b \\ d & e \end{vmatrix}\boldsymbol{k}$$

$$= (bf - ce)\,\mathbf{i} + (cd - af)\,\mathbf{j} + (ae - bd)\,\mathbf{k}$$

<u>Example</u>: Find the cross product $\mathbf{u} \times \mathbf{v}$ if

$$\mathbf{u} = 3\mathbf{i} + 2\mathbf{j} - 3\mathbf{k} \qquad \mathbf{v} = 3\mathbf{j} + 4\mathbf{k}$$

Solution

$$\begin{vmatrix} \boldsymbol{i} & \boldsymbol{j} & \boldsymbol{k} \\ 3 & 2 & -3 \\ 0 & 3 & 4 \end{vmatrix} = \begin{vmatrix} 2 & -3 \\ 3 & 4 \end{vmatrix}\boldsymbol{i} - \begin{vmatrix} 3 & -3 \\ 0 & 4 \end{vmatrix}\boldsymbol{j} + \begin{vmatrix} 3 & 2 \\ 0 & 3 \end{vmatrix}\boldsymbol{k}$$

$$= 17\boldsymbol{i} - 12\boldsymbol{j} + 9\boldsymbol{k}$$

The magnitude is the product of the lengths of the two vectors and the sine of the smaller angle between them measured from $\mathbf{u}$ to $\mathbf{v}$. Because of the "$\sin\theta$" factor, the cross product of the angles 0 and π (i.e., 180°) is zero. The cross product is anticommutative:

$$\mathbf{u} \times \mathbf{v} = -\mathbf{v} \times \mathbf{u}.$$

The cross product $\boldsymbol{a} \times \boldsymbol{b}$ is defined as a vector $\boldsymbol{c}$ that is perpendicular to both $\boldsymbol{a}$ and $\boldsymbol{b}$, with a direction given by the right-hand rule and a magnitude equal to the area of the parallelogram that the vectors span. The cross product is defined by the formula

$$\mathbf{a}\times\mathbf{b} = ||\mathbf{a}||\ ||\mathbf{b}||\sin\theta$$

where θ is the angle between ***a*** and ***b*** in the plane containing them (hence, it is between 0° and 180°), ‖***a***‖ and ‖***b***‖ are the magnitudes of vectors ***a*** and ***b***, and ***n*** is a unit vector perpendicular to the plane containing ***a*** and ***b*** in the direction given by the right-hand rule (illustrated to the right). If the vectors ***a*** and ***b*** are parallel (i.e., the angle θ between them is either 0° or 180°), by the above formula, the cross product of ***a*** and ***b*** is the zero vector **0**.

Cross product of unit vectors

For ***i*** × ***j*** the magnitude is 1 and the direction is ***k***, hence ***i*** × ***j*** = ***k***.

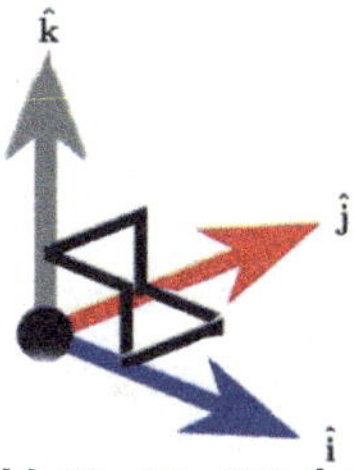

Also, ***j*** × ***k*** = ***i*** ***k*** × ***i*** = ***j*** and ***j*** × ***i*** = -***k*** ***k*** × ***j*** = -***i*** ***i*** × ***k*** = -***j*** .

Because ***j*** × ||***k***|| ||***j***|| ***cos*** θ *if* $\theta = 0°$ or $\theta = 180°$

then ***j*** × ***j*** = ***k*** × ***k*** = ***i*** × ***i*** = 0

Properties of the cross product

If **u**, **v**, and **w** are vectors and c is a scalar, then

1. **u** × **v** = - **v** × **u**
2. (c**u**) × **v** = c(**u** × **v**) = **u** × (c**v**)
3. **u** × (**v** + **w**) = **u** × **v** + **u** × **w**
4. (**u** + **v**) × **w** = **u** × **w** + **v** × **w**
5. **u**·(**v** × **w**) = (**u** × **v**)·**w**
6. **u** × (**v** × **w**) = (**u**·**w**)**v** - (**u**·**v**)**w**

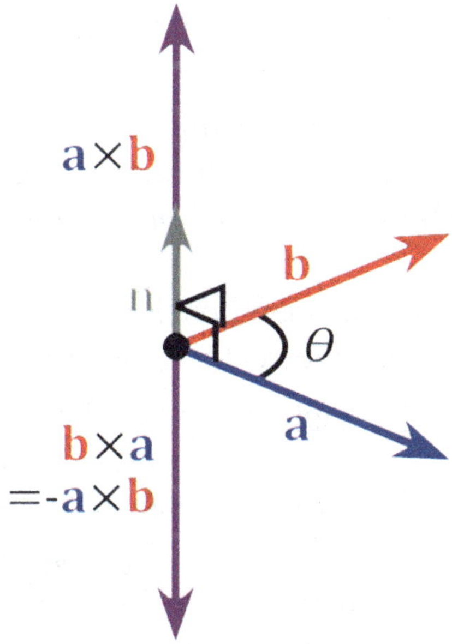

The cross product is not associative; that is **u** ×(**v** × **w**) ≠ (**u** × **v**) × **w**. This is easily proven with a counter-example (try using unit vectors).

Vectors in geometry

Example: Let **u** = < 2, 4 > and **v** = < 8, 4 > be vectors and consider the parallelogram that the two vectors make. Then ||**v** × **u**|| = Area of the Parallelogram and the direction of ***v*** × ***u*** is a right angle to the plane of the parallelogram (follows the right hand rule).

Proof: (See the figure on the next page)

From geometry we know that the area of a parallelogram is hb where h is the height (distance between two parallel sides) and b is the length of either of those sides. See the figure.

From trigonometry we know that the height shown is opposite to the angle θ and when divided by the magnitude of **u** is sin θ. Thus, h = ||**u**||sin θ. The area, hb, of the parallelogram, then is h||**v**||or ||**u**|| ||**v**||sin θ which is the definition of the cross product **u** × **v**.

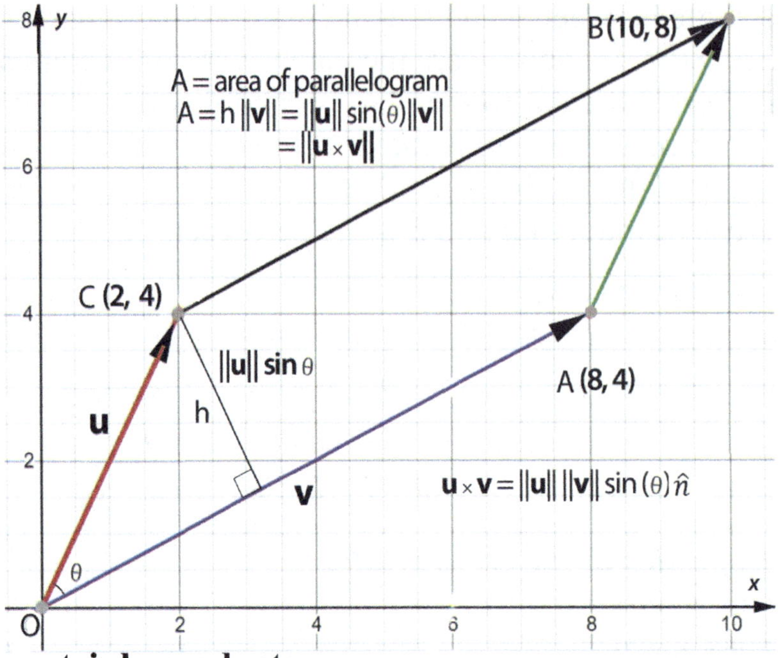

Three vectors - triple products

Let **u**, **v**, and **w**, be vectors. The *scalar triple product* of **u**, **v**, and **w** is defined as

$$\mathbf{u}\cdot(\mathbf{v}\times\mathbf{w}) = \mathbf{v}\cdot(\mathbf{w}\times\mathbf{u}) = \mathbf{w}\cdot(\mathbf{u}\times\mathbf{v}).$$

The absolute value of the scalar triple product is the volume of a parallelepiped defined by the three vectors.

The *vector triple product* is defined as

$$\mathbf{u}\times(\mathbf{v}\times\mathbf{w}) = (\mathbf{u}\cdot\mathbf{w})\,\mathbf{v} - (\mathbf{u}\cdot\mathbf{v})\,\mathbf{w}$$

Note the dot products in parentheses on the right side of the equation; since these are scalars, the vector triple product is the difference of two vectors and thus a vector.

<u>Example</u>: Calculate the scalar triple product and the vector triple product of

$$\mathbf{u} = \langle 3, 2, -3\rangle;\ \mathbf{v} = \langle 0, 3, 4\rangle;\ \mathbf{w} = \langle 2, 3, 4\rangle$$

Let's first calculate the cross products, (**u** × **v**), (**v** × **w**), and (**w** × **u**).

$$\mathbf{u}\times\mathbf{v} = \begin{vmatrix} \hat{x} & \hat{y} & \hat{z} \\ 3 & 2 & -3 \\ 0 & 3 & 4 \end{vmatrix} = \begin{vmatrix} 2 & -3 \\ 3 & 4 \end{vmatrix}\hat{x} - \begin{vmatrix} 3 & -3 \\ 0 & 4 \end{vmatrix}\hat{y} + \begin{vmatrix} 3 & 2 \\ 0 & 3 \end{vmatrix}\hat{z}$$

$$= (8+9)\,\hat{x} - (12-0)\,\hat{y} + (9-0)\,\hat{z}$$

$$= 17\hat{x} - 12\hat{y} + 9\hat{z}$$

$$= \langle 17, -12, 9\rangle$$

Similarly, (**v** × **w**) = <0, 8, -6> and (**w** × **u**) = <-17, 18, -5>

Now let's calculate (**u**·**w**) and (**u**·**v**):

u·**w** = <3, 2, -3> · <2, 3, 4> = 6 + 6 – 12 = 0 which means the vectors are orthogonal.

(**u**·**v**) = <3, 3, -3> · <0, 3, 4> = 0 + 9 –12 = -3

That was easy enough! Now to the triple products.

The *scalar triple product* of **u**, **v**, and **w** is $\mathbf{u}\cdot(\mathbf{v}\times\mathbf{w}) = \mathbf{v}\cdot(\mathbf{w}\times\mathbf{u}) = \mathbf{w}\cdot(\mathbf{u}\times\mathbf{v})$.

$$\mathbf{u}\cdot(\mathbf{v}\times\mathbf{w}) = <3, 2, -3> \cdot <0, 8, -6> = 0 + 16 + 18 = 34$$

Check: $\mathbf{u}\cdot(\mathbf{v}\times\mathbf{w}) = \mathbf{v}\cdot(\mathbf{w}\times\mathbf{u}) = <0, 3, 4> \cdot <-17, 18, -5> = 0 + 54 - 20 = 34$ ✓

The *vector triple product* is $\mathbf{u}\times(\mathbf{v}\times\mathbf{w}) = (\mathbf{u}\cdot\mathbf{w})\,\mathbf{v} - (\mathbf{u}\cdot\mathbf{v})\,\mathbf{w}$

$$= (0)<0, 3, 4> - (-6)<2, 3, 4> = <12, 18, 24>.$$

Parallelepipeds

The volume of a parallelepiped spanned by three vectors **u**, **v**, and **w**, is the triple product:

$$\text{Volume} = \mathbf{u}\cdot(\mathbf{v}\times\mathbf{w})$$

This can be found by computing the determinant of the three vectors:

$$\left|\begin{pmatrix} u_1 & v_1 & w_1 \\ u_2 & v_2 & w_2 \\ u_3 & v_3 & w_3 \end{pmatrix}\right|$$

The volume of a parallelepiped is the product of the area of its base A and its height h. The base is any of the six faces of the parallelepiped. The height is the perpendicular distance between the base and the opposite face. The vectors $\mathbf{x} = (x_1, x_2, x_3)$, $\mathbf{y} = (y_1, y_2, y_3)$, and $\mathbf{z} = (z_1, z_2, z_3)$ represent three edges that meet at one vertex. The volume of the parallelepiped then equals the absolute value of the scalar triple product

$$\mathbf{x}\cdot(\mathbf{y}\times\mathbf{z}):$$

$$V = \|\mathbf{x}\cdot(\mathbf{y}\times\mathbf{z})\| = \|\mathbf{y}\cdot(\mathbf{z}\times\mathbf{x})\| = \|\mathbf{z}\cdot(\mathbf{x}\times\mathbf{y})\|.$$

Each cross product is the area of a base.

$$A = \|\mathbf{x}\|\,\|\mathbf{y}\|\sin\theta = \|\mathbf{x}\times\mathbf{y}\|$$

where θ is the angle between **x** and **y**, and the height is

$$h = \|\mathbf{z}\|\cos\alpha$$

α is the internal angle between **z** and h.

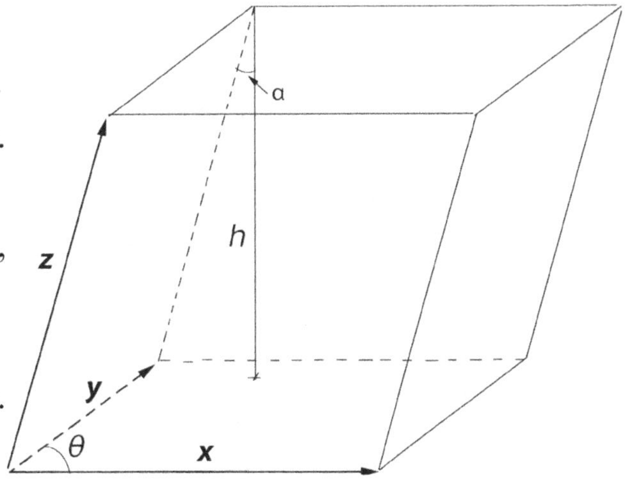

Other terms relating to vectors

The cross product of two vectors $\boldsymbol{a}$ and $\boldsymbol{b}$ is defined only in three-dimensional space and is denoted by $\boldsymbol{a}\times\boldsymbol{b}$. In physics, sometimes the notation $\boldsymbol{a}\wedge\boldsymbol{b}$ is used, though this is generally avoided in mathematics to avoid confusion with the exterior product, all beyond the scope of this book. Advanced linear algebra may define the *norm* (length) of a vector, *gradient* (∇), *curl* (or *divergence*), *vectorplot visualization*, and more - again, all beyond the scope of this book.

•

INDEX

Q

R

S

T

U

V

• • •

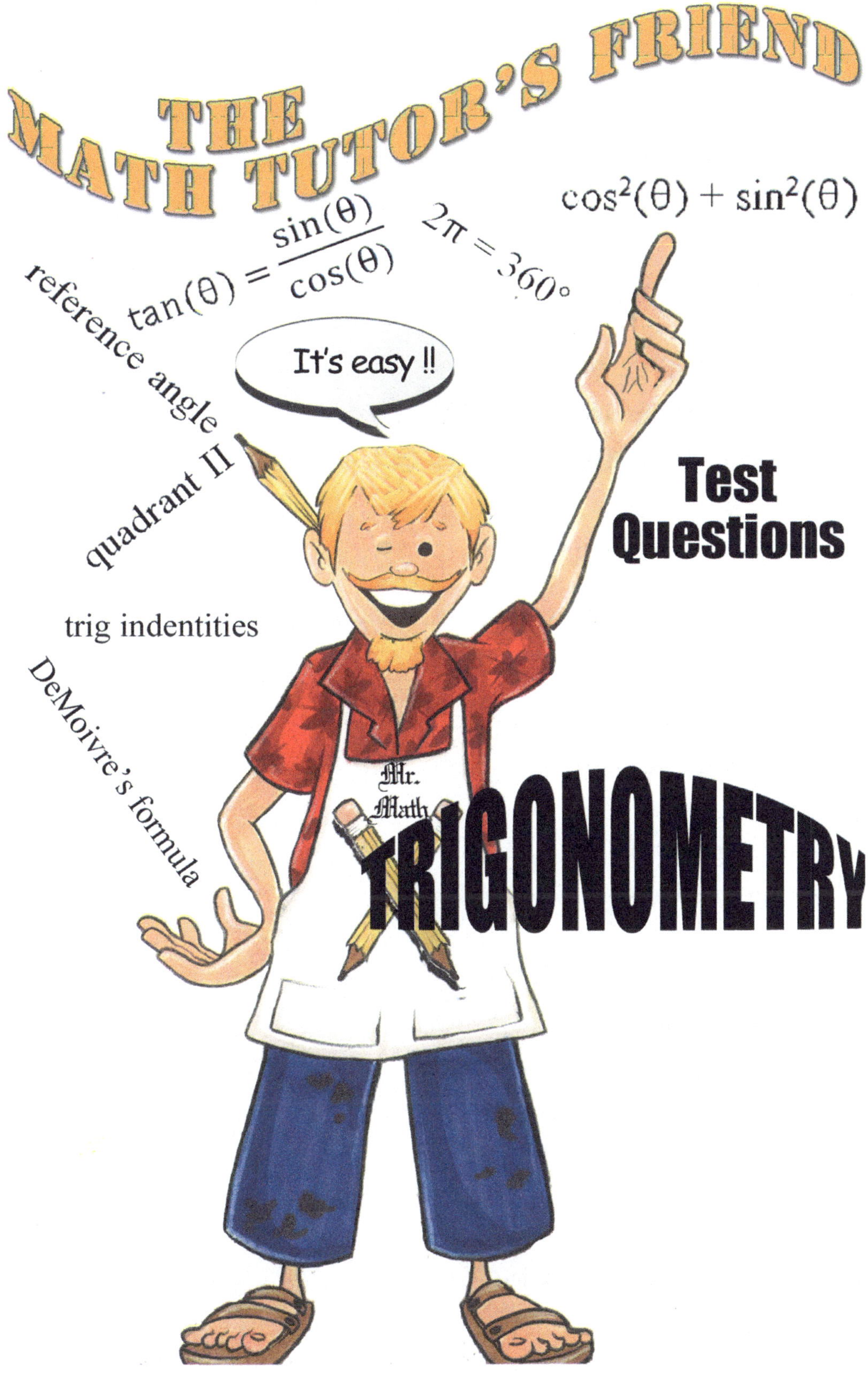
THE MATH TUTOR'S FRIEND
$\tan(\theta) = \frac{\sin(\theta)}{\cos(\theta)}$
$2\pi = 360°$
$\cos^2(\theta) + \sin^2(\theta)$
reference angle
It's easy !!
quadrant II
Test Questions
trig indentities
DeMoivre's formula
Mr. Math
TRIGONOMETRY

QUIZZES

•

• Angle AOB is in standard position - its vertex is

A. A and its terminal side lies on the x-axis

B. B and its terminal side lies on the y-axis

C. O which lies at the origin

D. All of the above

E. None of the above

• • Answer C.

An angle is said to be in *standard position* if its vertex is located at the origin of the x-y Cartesian coordinate plane and one ray is on the positive x-axis.

• ♥ •

• The Unit Circle has radius equal to

A. One

B. The trig ratio

C. Sine

D. Cosine

E. None of the above

• • Answer A.

• The trig sine ratio for vertex angle θ of a right triangle in standard position within the unit circle is

A. the length of the side opposite to θ

B. the length side adjacent to θ

C. the reciprocal of the length of the side opposite to θ

D. the reciprocal of the length of the side adjacent to θ

E. None of the above

• • Answer A.

The sine ratio is defined as the length of the opposite side divided by the length of the hypotenuse which is 1 in the unit circle. An angle is said to be in *standard position* if its vertex is located at the origin of the x-y Cartesian coordinate plane and one ray is on the positive x-axis.

• ♥ •

• The trig cosine ratio for vertex angle θ of a right triangle in standard position within the unit circle is

A. the length of the side opposite to θ

B. the length side adjacent to θ

C. the reciprocal of the length of the side opposite to θ

D. the reciprocal of the length of the side adjacent to θ

E. None of the above

• • Answer B.

The cosine ratio is defined as the length of the adjacent side divided by the length of the hypotenuse which is 1 in the unit circle. An angle is said to be in *standard position* if its vertex is located at the origin of the x-y Cartesian coordinate plane and one ray is on the positive x-axis.

• The trig tangent ratio for vertex angle θ of a right triangle in standard position within the unit circle is

A. the length of the side opposite to θ

B. the length side adjacent to θ

C. the reciprocal of the length of the side opposite to θ

D. the reciprocal of the length of the side adjacent to θ

E. None of the above

• • Answer E. The tangent ratio is defined as the length of the opposite side divided by the length of the adjacent side in the unit circle. An angle is said to be in *standard position* if its vertex is located at the origin of the *x-y* Cartesian coordinate plane and one ray is on the positive *x*-axis.

• ♥ •

• The trig secant ratio for vertex angle θ of a right triangle in standard position within the unit circle is

A. the length of the side opposite to θ

B. the length side adjacent to θ

C. the reciprocal of the length of the side opposite to θ

D. the reciprocal of the length of the side adjacent to θ

E. None of the above

• • Answer D. The secant ratio is the reciprocal of the cosine. An angle is said to be in *standard position* if its vertex is located at the origin of the *x-y* Cartesian coordinate plane and one ray is on the positive *x*-axis.

• The signs of the cosine and sine in Quadrant I of the Unit Circle are, respectively,

A. + and - (plus and minus)

B. + and + (plus and plus)

C. - and + (minus and plus)

D. - and - (minus and minus)

E. None of the above

• • Answer B

• ♥ •

• The signs of the cosine and sine in Quadrant II of the Unit Circle are, respectively,

A. + and - (plus and minus)

B. + and + (plus and plus)

C. - and + (minus and plus)

D. - and - (minus and minus)

E. None of the above

• • Answer C

• The signs of the cosine and sine in Quadrant III of the Unit Circle are, respectively,

A. + and - (plus and minus)

B. + and + (plus and plus)

C. - and + (minus and plus)

D. - and - (minus and minus)

E. None of the above

• • Answer D

• ♥ •

• The signs of the cosine and sine in Quadrant IV of the Unit Circle are, respectively,

A. + and - (plus and minus)

B. + and + (plus and plus)

C. - and + (minus and plus)

D. - and - (minus and minus)

E. None of the above

• • Answer A

• The signs of the tangent and cotangent in Quadrant I of the Unit Circle are, respectively,

A. + and + (plus and plus)

B. + and - (plus and minus)

C. - and + (minus and plus)

D. - and - (minus and minus)

E. None of the above

• • Answer A

• ♥ •

• The signs of the tangent and cotangent in Quadrant II of the Unit Circle are, respectively,

A. + and + (plus and plus)

B. + and - (plus and minus)

C. - and + (minus and plus)

D. - and - (minus and minus)

E. None of the above

• • Answer D

• The signs of the tangent and cotangent in Quadrant III of the Unit Circle are, respectively,

A. + and + (plus and plus)

B. + and - (plus and minus)

C. - and + (minus and plus)

D. - and - (minus and minus)

E. None of the above

• • Answer A

• ♥ •

• The signs of the tangent and cotangent in Quadrant IV of the Unit Circle are, respectively,

A. + and + (plus and plus)

B. + and - (plus and minus)

C. - and + (minus and plus)

D. - and - (minus and minus)

E. None of the above

• • Answer D

• An angle of $-5\pi/6$ on the Unit Circle has a reference angle of

A. $-5\pi/6$

B. $5\pi/6$

C. $\pi - \pi/6$

D. $\pi/6$

E. None of the above

• • Answer D.

The reference angle is the acute angle measured from the terminal side of a given angle to the x-axis. An angle of $-5\pi/6$ is $\pi/6$ in Quadrant III.

• ♥ •

• The reference angle for $\pi/4$ is

A. $-\pi/4$

B. $3\pi/4$

C. $-3\pi/4$

D. $\pi/4$

E. None of the above

• • Answer D.

The angle $\pi/4$ lies in Quadrant I, and is, therefore, its own reference angle.

• With respect to the Unit Circle, a reference angle is measured from the terminal side of a given angle to the

A. y-axis

B. x-axis

C. y-axis – π

D. x-axis + π

E. None of the above

• • Answer B.

A reference angle is always measured to the x-axis nearest to the terminal side.

• ♥ •

• The angle 20° is _____ in radian measure.

A. π/3

B. π/6

C. π/9

D. π/12

E. None of the above

• • Answer C.

The proportion is $\boldsymbol{d}/180 = \boldsymbol{r}/\pi$. In this problem, the unknown, $\boldsymbol{r}$, is calculated as follows: $\boldsymbol{r} = \pi\boldsymbol{d}/180$. Substituting $\boldsymbol{d} = 20°$, $\boldsymbol{r} = 20\pi/180 = \pi/9$.

Alternatively, from the definition π radians = 180 degrees, divide 180 by 9 to get 20 degrees and divide π by 9 also.

• $cos(x)(1 + sec(x)) =$

A. 1
B. $tan(x)$
C. $1 + cos(x)$.
D. $1 - sin(x)$
E. None of the above

• • Answer C.
Distribute $cos(x)$ to get

$$cos(x) + cos(x)sec(x) = cos(x) + cos(x)/\,cos(x) = cos(x) + 1$$

• ♥ •

$(1 - cos^2(x))(1 + cot^2(x)) =$

A. 1
B. $tan(x)$
C. $1 + cos(x)$.
D. $1 - sin(x)$
E. None of the above

• • Answer A.
Noting that $\sin^2(x) + \cos^2(x) = 1$,

$$(1 - cos^2(x))(1 + cot^2(x)) = sin^2(x)(1 + cot^2(x))$$
$$= sin^2(x) + sin^2(x)cos^2(x)/sin^2(x)$$
$$= sin^2(x) + cos^2(x) = 1$$

• The base angle A of an isosceles triangle ***ABC*** is 30° with adjacent side ***AC*** of length 5.

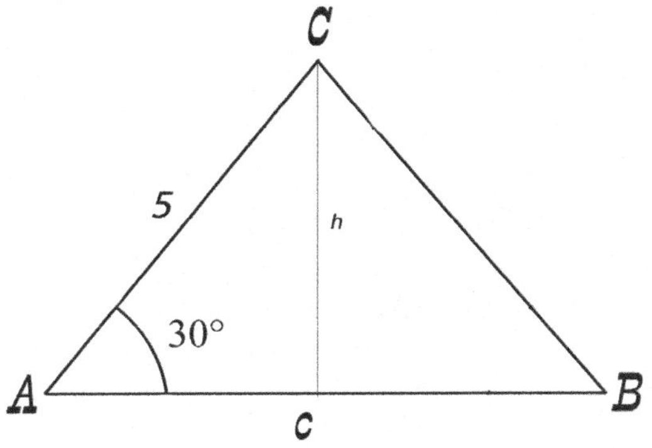

a. What is the perimeter of the triangle ?

b. What is the height of the triangle?

(round to tenths)

A. 15.7 and 2.5

B. 18.7 and 4.3

C. 14.7 and 2.5

D. 8.7 and 4.3

E. None of the above

• • Answer B.

a. Use the Law of Sines.

$$\sin(A)/a = \sin(C)/c$$

Because the triangle is isosceles, each base angle (A and B) is 30° and the vertex angle, C, is $180° - 2\times30° = 120°$. The length of c is the unknown and both sides adjacent to the vertex angle are of length 5.

Substituting given values, we have

$\sin(30°)/5 = \sin(120°)/c$

from which $c = 8.7$ to one decimal place. The perimeter is

$$5 + 5 + 8.7 = 18.7$$

b. $h = \sin(30°) = (\sqrt{3})/2 \approx 4.3$

• The length **a** (to the nearest tenth) is

A. 4.1

B. 7.2

C. 5.7

D. 9.1

E. None of the above

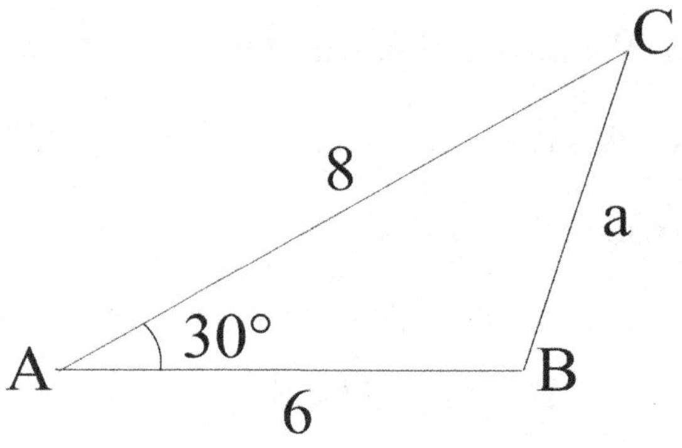

• • Answer A.

Use the Law of Cosines: $\mathbf{a}^2 = 8^2 + 6^2 - 2 \times 6 \times 8 \times \cos(30°)$

$= 64 + 36 - 83.1 \approx 16.9$

$\sqrt{16.9} \approx 4.1$

• ♥ •

• $[\cos(\pi/6) + \boldsymbol{i} \sin(\pi/6)]^{12} =$

A. -1

B. 1

C. 0

D. $1 - \boldsymbol{i}$

E. None of the above

• • Answer B.

Use DeMoivre's Theorem

$$[\cos(\pi/6) + \boldsymbol{i} \sin(\pi/6)]^{12} = \cos(12\pi/6) + \boldsymbol{i} \sin(12\pi/6)$$

$$= \cos(2\pi) + \boldsymbol{i} \sin(2\pi) = 1 + \boldsymbol{i}\, 0 = 1$$

• $(cos(\pi/8) + \boldsymbol{i}sin(\pi/8))^4 =$

A. $\boldsymbol{i}$
B. 1
C. 0
D. 1/2
E. None of the above

• • Answer A.
Use DeMoivre's Theorem

$$(cos(\pi/8) + \boldsymbol{i}sin(\pi/8))^4 = cos(4\pi/8) + \boldsymbol{i}sin(4\pi/8)$$
$$= cos(\pi/2) + \boldsymbol{i}sin(\pi/2) = 0 + \boldsymbol{i} \times 1 = \boldsymbol{i}$$

• ♥ •

• $sin^4(x) - cos^4(x) =$

A. $(sin(x) - \cos(x))(sin(x) + cos(x))$
B. $\sin(4x)$
C. $\cos(4x)$
D. $sin^2(x) + cos^2(x)$.

• • Answer A.
This is a "difference of squares" pattern.

$$sin^4(x) - cos^4(x) = (sin^2(x) - cos^2(x))(sin^2(x) + cos^2(x))$$

The term $(sin^2(x) + cos^2(x))$ is simply the familiar Pythagorean identity, and equals 1; the term $(sin^2(x) - cos^2(x))$ is also a difference of two squares, so

$$sin^4(x) - cos^4(x) = (sin^2(x) - cos^2(x))(1)$$
$$= (sin(x) - \cos(x))(sin(x) + cos(x)).$$

• One of the “standard” triangle is the so-called 3-4-5 triangle (others include the “45° triangle” and the “30°-60°-90° triangle”). What are the three angles of the 3-4-5 triangle to one decimal place?

A. 30.0° 60.0° 90.0°

B. 19.9° 40.1° 50.9°

C. 36.9° 53.1° 90.0°

D. 42.8° 47.2° 90.0°

• • Answer C. Use the Law of Cosines:

Let the three sides be $a = 3$; $b = 4$; $c = 5$ and the opposite angles A, B. and C, respectively. From the Law of Cosines,

$C = \cos^{-1}((a^2 + b^2 - c^2)/2ab) = \cos^{-1}(9 + 16 - 25)/50) = 90.0°$

$B = \cos^{-1}((9 + 25 - 16)/30) = \cos^{-1}(3/5) = 53.1°$

and since the internal angles of a triangle sum to 180°

$A = 180° - (A + B) = 36.9°$

• ♥ •

• A “30°-60°-90° triangle” has a hypotenuse measuring 12 inches. What is the length of the shortest leg of the triangle?

A. 3 inches

B. 4 inches

C. 6 inches

D. 8 inches

E. None of the above

• • Answer C.

The shortest leg is opposite the smallest angle (30°). The opposite leg is $\sin(30°)\times 12$ inches = 6 inches.

• Find the volume of the parallelepiped spanned by the vectors

u = < 1, 0, 2 >   v = < 0, 2, 3 >   w = < 0, 1, 3 >

A. 3

B. 4

C. 5

D. 6

E. None of the above

• • Answer A.

Take the determinant of the vectors set up in a 3×3 matrix:

$$\left|\begin{pmatrix} 1 & 0 & 2 \\ 0 & 2 & 3 \\ 0 & 1 & 3 \end{pmatrix}\right| = 3$$

• ♥ •

• If the angle between **u** and **v** is 36.9°, then the area of the parallelogram is approximately

A. 16

B. 24

C. 32

D. 40

E. None of the above

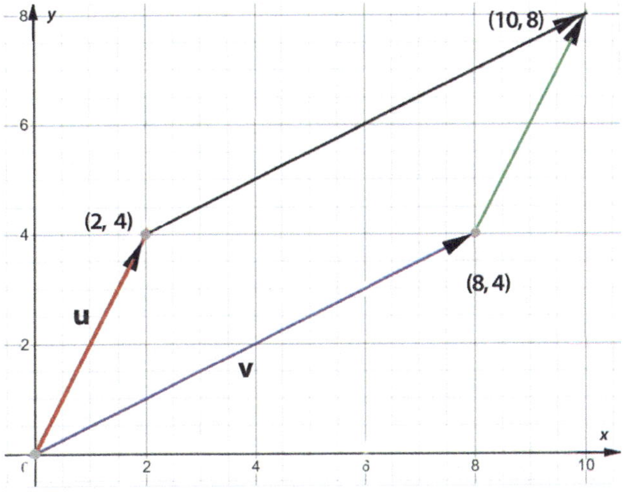

• • Answer B.

u = < 2, 4 > **v** = < 8, 4 >

|| **u** || = **2√5** || **v** || = 4√5

The cross product **u** × **v** = || **u** || || **v** || sin (36.9°)

= 40sin(36.9°) = 24.02, approximately

• Convert the vector **v** = < 6, 8 > to a unit vector.

A. < 0.6, 0.8 >

B. < 1, 1 >

C. < 8, 6 >

D. < 3/7, 4/7 >

E. None of the above

• •Answer A.

$\| \mathbf{v} \| = \sqrt{(6^2 + 8^2)} = 10$

$\mathbf{v}/\| \mathbf{v} \| = \langle 0.6, 0.8 \rangle$

• ♥ •

• In the **i, j, k** coordinate system, **j**×**k** =

A. **i**

B. **j**

C. **k**

D. 1

E. None of the above

• • Answer A.

Use the right-hand rule

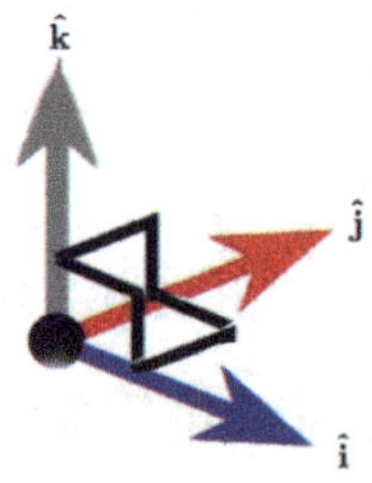

• What is the dot product of the vector **u** with itself (i.e., **u**·**u** =)?

A. 2**u**

B. 2·(**u** × **u**)

C. 2|| **u** ||

D. || **u** ||^2

E. None of the above

• • Answer D.

This is one of the properties of the dot product.

• ♥ •

• The angle between the vectors **u** = < 2, 3 > and **v** = < 4, 1 > is (round your answer to the nearest degree)

A. 11°

B. 31°

C. 42°

D. 73°

E. None of the above

• • Answer C.

The dot product is **u**·**v** = (2)(4) + (3)(1) = 11

|| **u** ||^2 = $2^2 + 3^2$ = 13 so || **u** || = √13; || **v** ||^2 = $4^2 + 1^2$ = 17

so || **v** || = √17

cos θ = **u**·**v**/(|| **u** || || **v** ||)

finally, θ = arccosine(11/(√13 √17)) ≈ 42°

• Calculate **u** × **v** where **u** = < 3, 2, -3 >; **v** = < 0, 3, 4 >; **w** = < 2, 3, 4 >

A. < 0, 8, -6 >
B. < -17, 198, -5 >
C. < 17, -12, 9 >
D. All of the above
E. None of the above

• • Answer C.

$$\mathbf{u} \times \mathbf{v} = \begin{vmatrix} \hat{x} & \hat{y} & \hat{z} \\ 3 & 2 & -3 \\ 0 & 3 & 4 \end{vmatrix} = \begin{vmatrix} 2 & -3 \\ 3 & 4 \end{vmatrix} \hat{x} - \begin{vmatrix} 3 & -3 \\ 0 & 4 \end{vmatrix} \hat{y} + \begin{vmatrix} 3 & 2 \\ 0 & 3 \end{vmatrix} \hat{z}$$

$$= (8 + 9)\,\hat{x} - (12 - 0)\,\hat{y} + (9 - 0)\,\hat{z}$$

$$= 17\hat{x} - 12\hat{y} + 9\hat{z}$$

$$= \langle 17, -12, 9 \rangle$$

• ♥ •

• Calculate (**u**·**w**): where **u** = < 3, 2, -3 >; **v** = < 0, 3, 4 >; **w** = < 2, 3, 4 >

A. < 0, 6, -12 >

B. 3

C. -3

D. 0

E. None of the above

• • Answer D.

u·**w** = < 3, 2, -3 > ·< 2, 3, 4 > = 6 + 6 – 12 = 0 which means the vectors are orthogonal.

• If the dot product of the non-zero vectors **a** and **b** is zero then the angle between them is

A. $\pi/2$

B. $\pi/3$

C. $\pi/4$

D. $\pi/6$

E. None of the above

• • Answer A.

$$\mathbf{a}\cdot\mathbf{b} = ||\mathbf{a}||\ ||\mathbf{b}||\cos(\pi/2) = (||\mathbf{a}||\ ||\mathbf{b}||)(0) = 0$$

• ♥ •

• If two non-zero vectors **a** and **b** are parallel then their dot product is

A. Undefined

B. 0

C. $\pi/2$

D. π

E. None of the above

• • Answer E.

$$\mathbf{a}\cdot\mathbf{b} = ||\mathbf{a}||\ ||\mathbf{b}||\cos(0) = (||\mathbf{a}||\ ||\mathbf{b}||)(1) = ||\mathbf{a}||\ ||\mathbf{b}||$$

• DeMoivre's Formula

A. is the same as Euler's Formula

B. can be used to find powers of cis(θ)

C. is not employed in trigonometry

D. is the solution to $\sin^2(\theta) + \cos^2(\theta)$

E. None of the above

• • Answer B

• ♥ •

• An angle in standard position

A. has its initial side lying on the x-axis

B. has its terminal side lying on the y-axis

C. is 90° (π/2)

D. has no vertex

E. None of the above

• • Answer A

• The reference angle for $\theta = 390°$ is

A. 390°

B. 90°

C. 30°

D. -90°

E. None of the above

• • Answer C.

A reference angle is the positive acute angle relative to the *x*-axis.

• ♥ •

• The reference angle for $\theta = -60°$ is

A. 360°

B. 60°

C. 300°

D. -30°

E. None of the above

• • Answer B.

The Math Tutor's Friend

Informal Review of High-School Math Topics for Tutors

-7-

• Trigonometry •